ROUTLEDGE LIBRARY EDITIONS:
COLD WAR SECURITY STUDIES

Volume 60

THE USSR AND THE WESTERN ALLIANCE

THE USSR AND THE WESTERN ALLIANCE

Edited by
ROBBIN F. LAIRD AND SUSAN L. CLARK

LONDON AND NEW YORK

First published in 1990 by Unwin Hyman, Inc.

This edition first published in 2021
by Routledge
2 Park Square, Milton Park, Abingdon, Oxon OX14 4RN

and by Routledge
52 Vanderbilt Avenue, New York, NY 10017

Routledge is an imprint of the Taylor & Francis Group, an informa business

© 1990 Unwin Hyman, Inc.

All rights reserved. No part of this book may be reprinted or reproduced or utilised in any form or by any electronic, mechanical, or other means, now known or hereafter invented, including photocopying and recording, or in any information storage or retrieval system, without permission in writing from the publishers.

Trademark notice: Product or corporate names may be trademarks or registered trademarks, and are used only for identification and explanation without intent to infringe.

British Library Cataloguing in Publication Data
A catalogue record for this book is available from the British Library

ISBN: 978-0-367-56630-2 (Set)
ISBN: 978-1-00-312438-2 (Set) (ebk)
ISBN: 978-0-367-56831-3 (Volume 60) (hbk)
ISBN: 978-0-367-56832-0 (Volume 60) (pbk)
ISBN: 978-1-00-309950-5 (Volume 60) (ebk)

Publisher's Note
The publisher has gone to great lengths to ensure the quality of this reprint but points out that some imperfections in the original copies may be apparent.

Disclaimer
The publisher has made every effort to trace copyright holders and would welcome correspondence from those they have been unable to trace.

The USSR and the Western Alliance

Robbin F. Laird & Susan L. Clark

With contributions by:

**Hannes Adomeit
Charles Gati
Erik P. Hoffmann
Phillip A. Petersen
Notra Trulock III**

Boston
Unwin Hyman
London   Sydney   Wellington

© 1990 by Unwin Hyman, Inc.
This book is copyright under the Berne Convention. No
reproduction without permission. All rights reserved.

Unwin Hyman, Inc.
8 Winchester Place
Winchester, Massachusetts 01890, USA

Published by the Academic Division of
Unwin Hyman Ltd
15/17 Broadwick Street
London W1V 1FP, UK

Allen & Unwin (Australia) Ltd
8 Napier Street
North Sydney, NSW 2060, Australia

Allen & Unwin (New Zealand) Ltd in association with the
Port Nicholson Press Ltd
Compusales Building, 75 Ghuznee Street
Wellington 1, New Zealand

First published in 1990.

Library of Congress Cataloging-in-Publication Data

Laird, Robbin F. (Robbin Frederick), 1946-
　The USSR and the western alliance / Robbin F. Laird and Susan L.
Clark ; with contributions by Hannes Adomeit ... [et al.].
　　p.　　cm.
　Includes index
　ISBN 0-04-445392-2
　1. North Atlantic Treaty Organization. 2. Soviet Union—Military
policy. 3. Europe—Military policy. 4. Arms control. I. Clark,
Susan L., 1959- . II. Adomeit, Hannes. III. Title.
UA646.3.L32 1989
355'.033'048—dc20　　　　　　　　　　　　　　　　　　　　　　　89-33761
　　　　　　　　　　　　　　　　　　　　　　　　　　　　　　　　CIP

British Library Cataloguing in Publication Data

　　The USSR and the Western alliance.
　　1. North Atlantic region. International
security. Implications of military policies of
the Soviet Union.
　　2. Soviet Union. Military policies.
Implications for international security of
North Atlantic region
I. Clark, Susan II. Laird, Robbin F.
(Robbin Frederick), *1946–* III. Adomeit,
Hannes
327.1'16

　　ISBN 0–04–445392–2

Typeset in 10 on 12 point Garamond and printed in Great Britain by
The University Press, Cambridge

To Erik and Lauren

Contents

Abbreviations		*page* ix
Acknowledgments		xi
1	The Soviet Union and the Western Alliance *Robbin F. Laird*	1
2	Soviet Views and Policies toward Theater War in Europe *Phillip A. Petersen and Notra Trulock III*	25
3	Gorbachev and the Western Alliance: Reassessing the Anticoalition Strategy *Erik P. Hoffmann*	51
4	Eastern Europe as a Factor in Soviet Foreign Policy toward Western Europe *Charles Gati*	89
5	Soviet Perspectives on French Security Policy *Robbin F. Laird*	105
6	Soviet Perspectives on British Security Policy *Robbin F. Laird and Susan L. Clark*	127
7	A Nuclear-Free Zone in Northern Europe *Susan L. Clark*	163
8	Soviet Security Perspectives on Germany *Hannes Adomeit*	181
9	Soviet Public Diplomacy toward West Germany under Gorbachev *Robbin F. Laird*	217
10	Soviet Challenges to NATO *Robbin F. Laird and Susan L. Clark*	251
About the Authors		257
Index		259

Abbreviations

FBIS: Foreign Broadcast Information Service, Daily Report: Soviet Union
FBIS FPD: Foreign Broadcast Information Service, Foreign Press Digest
FBIS-WE: Foreign Broadcast Information Service, Daily Report: Western Europe
IA: International Affairs (Moscow)
JPRS-UFM: Joint Publications Research Service, USSR Report: Foreign Military Review
JPRS-UMA: Joint Publications Research Service, USSR Report: Military Affairs
JPRS-WER: Joint Publications Research Service, West Europe Report
KVS: Kommunist Vooruzhennkh Sil
KZ: Krasnaya zvezda
MEMO: Mirovaya ekonomika i mezhdunarodnye otnosheniya
MZ: Mezhdunarodnaya zhizn'
RFE: Radio Free Europe Research
RL: Radio Liberty Research Bulletin
SShA: SShA: ekonomika, politika, ideologiya
SVE: Sovetskaya voennaya entsiklopediya
SWB: BBC Monitoring, Summary of World Broadcasts. Part 1: USSR
VIZ: Voenno-istoricheskii zhurnal
VM: Voennaya mysl'
VV: Voennyi vestnik
ZVO: Zarubezhnoe voennoe obozrenie

Acknowledgments

We would like to thank the many people who have helped bring this book to fruition. Specifically, we are indebted to our colleagues at the Institute for Defense Analyses for their administrative support. We would also like to thank those people both in the United States and in various West European countries who have contributed to our thinking about Soviet security policy. Finally, we are particularly appreciative of the many valuable insights provided by the IDA working group participants during our various seminars on the Soviet Union.

1
The Soviet Union and the Western Alliance

Robbin F. Laird
Institute for Defense Analyses

This book assesses Soviet analyses and perceptions of the nature of the Western Alliance, particularly those elements of its security policy most directly relevant to the military challenge. Within this broad framework, the security policies of the key West European members of the Western Alliance are examined and the East European factor in Soviet policy toward the Alliance is assessed. Although economic strategy is an important dimension of the Soviet approach to the Alliance, as is the evolution of American military policy, this book does not address these dimensions because they have been examined extensively elsewhere.

Two other critical dimensions of the Soviet approach to the Alliance have been examined as well: the public diplomacy component of Soviet policy and the political dimension of Soviet military strategy—specifically, the pursuit of an anticoalition strategy. It is in the public diplomacy arena that Soviet General Secretary Mikhail Gorbachev has introduced the greatest changes in Soviet strategy. In terms of military strategy, attention is focused on how Soviet military theory and practice for theater war in Europe have evolved over the past two decades. In this context, the centrality of political factors to the initial period of a crisis and war in Europe is explored. This introductory chapter examines all of these issues in general terms as a means of assessing the overall Soviet effort vis-à-vis the Western Alliance. The following chapters provide greater detail on each of the given topics.

The Gorbachev leadership is paying more rather than less attention to the place of politics in the evolution of Soviet military strategy. The oft-repeated notion that the Soviets are focusing more on the political dimension of Soviet strategy than on the military dimension simply means that their approach toward the Alliance has been altered. The balance between the political-military and the technical-military aspects is what is in question, not the salience of politics to Soviet military strategy.

General Soviet Perspectives on the Alliance

From the Soviet perspective, "Atlanticism" has allowed the United States to exercise considerable influence over political and economic developments within Western Europe.[1] American definitions of Western security needs dominated Western Europe's security interests as well as the latter's political and economic relationships with the Soviet Union. Such an Atlanticized Europe was able to exist only under the specific conditions of American predominance over Western Europe that existed in the late 1940s and early 1950s.

This type of Atlanticism has been seriously undercut by the changing balance of power (in Soviet terminology, the "shift in the correlation of forces") between the United States and Western Europe. The growth of West European economic might has led to the emergence of the West European power center in the capitalist world, a power center that has exercised growing assertiveness and even independence from the United States.

The resurgence of the West European power center has occurred most dramatically in the political-economic realm. Economically, Western Europe has become a serious competitor as well as a collaborator with the United States. West European scientific and technological capability is a significant factor in European economic development and also provides the basis for significant progress in cutting-edge industries. The West Europeans have become major players in trade and capital transfer relations with the Third World, in many ways having even more diversified economic relations with these countries than does the United States. Politically, the West Europeans have collaborated to an unprecedented extent in the creation and development of intra-European economic relations. Strong Soviet skepticism about the European Economic Community (EEC) has slowly given way to grudging recognition of the significance of transnational European cooperation. Notably, the Soviets are paying increasing attention to the process of the economic transformation of European markets associated with the 1992 target date for eliminating the remaining economic barriers in the Common Market.

The emergence of the West European power center has led to greater assertiveness within West-West military security relations as well. The West Europeans have expanded the scope and extent of intra-European security cooperation, most noticeably in armaments development and production. Cooperative efforts among the major players in European security (for example, between France and West Germany) have sought to increase Europe's weight in Alliance policy. One result has been the development of the West Europeans' own policies toward Eastern Europe and the Soviet Union.

Nonetheless, the West European effort to alter the balance of military security relations with the United States is limited by the fact that Western Europe is only a quasi–power center in military matters. The United States continues to possess the West's most significant nuclear component. It also remains the critical rear of the Alliance, which is vital to the reinforcement of the North Atlantic Treaty Organization's (NATO) forces in times of crisis. Furthermore, the United States possesses the West's major out-of-area military capability (only France and Britain remain key players, with Italy playing a growing role). It is simply not the case that West Europeans believe they can go it alone. Rather, they seek to define more effectively what is in their interest to do, both within Europe and within the U.S.-European relationship.

In short, a major shift has occurred within the Alliance. Although the United States remains the most significant military power, the West European countries are playing an increasingly critical role in the evolution of Western security policy. The United States may initiate changes in Western policy, but Europe can veto them. In this connection, West European economic and political development is a critical determinant of Western military capabilities.

Soviet analysts express concern that a stronger Western Europe, one more independent and more capable of defending itself, might be emerging. The Soviets hope to promote, or at least contribute to, a crisis of statecraft in the West and thereby impede the development of a more "mature" partnership between Western Europe and the United States. The Soviets hope to contribute to Atlanticism's decline without encouraging the further development of intra-European cooperation in economic, political, and military areas. They are especially anxious to impede the emergence of a better division of labor between Western Europe and the United States in security issues, a division that would allow the Americans to confront the Soviet Union more effectively both within and outside Europe.

The Soviet leadership has long held the position that the USSR can and should influence the United States's West European allies, which in turn can and should influence the United States. While the Soviets would clearly welcome a reduction in U.S. influence in Western and Eastern Europe, it has been implicitly acknowledged that the goal of eliminating U.S. influence in

Western Europe is neither feasible nor desirable.[2] Instead, West European nations must be induced to use more fully their potential to influence U.S. policy on key security, commercial, and scientific-technological issues.

The Gorbachev administration has placed less emphasis on the ideological differences between the "socialist" and "capitalist" nations of Europe and more emphasis on their geographic, cultural, and political bonds. In this connection, the theme of a "common European home"—a commonality of interests among the Soviet Union, Eastern Europe, and Western Europe—has been accorded a significant amount of attention. At the same time, Soviet leaders and commentators have also underscored the differences between U.S. and European interests.

Finally, since the October 1986 Reykjavik summit between General Secretary Gorbachev and President Reagan, Soviet analysts have been especially keen to counterpose the opportunities for East-West cooperation inherent in the theme of the "common European home" with the dangers evident in the more conservative manifestations of pronuclearism among key West European governments. As military commentator Vladimir Chernyshev noted:

> The prospects for the deliverance of Europe and the entire world from nuclear weapons, the possibility of which was shown by the meeting in Reykjavik, horrified some people in Western European capitals. In London, Bonn, and Paris they began to assert that nuclear weapons were almost a boon, that there was no alternative to nuclear strategic weapons, and that the elimination of U.S. nuclear arms in Europe would weaken the political and security ties between the United States and Western Europe.[3]

Especially of concern to the Soviets is the reverse effect of conservative West European elements, causing Washington to resist the "new" opportunities inherent in Gorbachev's foreign policy. For example, some Soviet analysts believe that conservative West European governments acted independently and successfully to influence U.S. policy in negative ways after Reykjavik.[4]

In short, in the environment that has evolved since the signing of the intermediate-range nuclear force (INF) treaty, the Soviets seek to develop a policy that influences the Americans to pursue arms control, constrains West European defense cooperation, and impedes other trends in Western Europe from influencing the Americans negatively.

Soviet Perspectives on Western Europe

Soviet peacetime strategy toward the Alliance is designed to test the Alliance framework, to search for weaknesses, and to encourage positive trends. The Soviets seek to deepen the fissures between the American and European components of the Alliance as well as to play on the pressure spots in intra-European relations.

Especially significant to Soviet peacetime strategy toward the Alliance has been the need to influence the shape and direction of West European foreign policy as the West struggles to define its East-West policy. Western Europe's increased assertiveness has provided the Soviets with opportunities to undercut U.S. influence in Western Europe and to hinder the development of various Western relationships (such as the strengthening of the Bonn-Paris relationship) that the Soviets find damaging to their interests.

A basic challenge for the Soviets is to manage as successfully as possible the mediated nature of their relationship with Western Europe. The Soviet–West European relationship is determined partly by the attempt to influence one another's allies. The Soviet leadership has sought to influence American policy toward the Soviet Union and to limit American foreign policy capabilities by molding West European foreign policy behavior. In turn, the West European powers have made a significant effort to shape Soviet foreign policy behavior by increasing West European ties with Eastern Europe, especially through trade and other forms of commercial relations.

Soviet assessments of the evolution of security policy in the key West European countries reflect a keen awareness of the shifting correlation of forces in Western relationships and the increasingly assertive role of Western Europe within the Alliance. These assessments underscore the necessity of crafting a more effective policy toward Western Europe both to influence the independent evolution of Western Europe and to shape American options and opportunities. Within this framework, West Germany remains the key to such Soviet efforts.

Soviet Policy toward West Germany

Throughout the postwar period, the Soviets have pursued various policies toward West Germany. The basic Soviet effort to influence the evolution of West German policy has wavered between military pressure and diplomatic persuasion, the former playing an increasingly important role beginning in the second half of the 1970s. The three Soviet leaders prior to Gorbachev apparently thought that forcing the pace of military competition in Europe and improving the Soviet preponderance in conventional and INF systems would make it possible to translate military power into political influence. The role of "nuclear hostage" was to be impressed on the West German mind as a way of lessening the West German attachment to the United States. Moscow viewed the rise of a powerful peace movement in West Germany as one of the main indicators that this approach was working.

What Soviet leaders until Gorbachev did not appreciate was that increased Soviet pressure produced counteractions. For example, this pressure, which raised international tensions, contributed to greater U.S.-West German cohesion. In addition, in the early 1980s, the balance among West German political parties shifted toward a more anti-Soviet stance, a shift that was caused partly by Soviet intransigence on the INF issue.

Soviet political-military analysts of the Federal Republic of Germany (FRG) understand that the "forward defense" of West Germany is the key element of its political-military strategy. The stationing of foreign troops on West German soil is legitimized as being to defend West German territorial integrity from the outset of armed hostilities. The commitment to nuclear deterrence is ambivalent in the FRG; only nuclear use outside West German territory is perceived to be legitimate, and the nuclear hostage fear is deeply rooted. Finally, West Germany's commitment to conventional defense puts it at odds with the other two major European powers (France and Britain), which rely on nuclear deterrence to defend their territory.

The political element of Soviet strategy takes into account these main elements of specific West German interests. With regard to forward defense, the Soviets link such defense with offensive operations. The growing Soviet effort to paint their doctrine in defensive terms is designed partly to undercut the legitimacy of forward defense in West German minds. As if to underline a direct parallel with the Nazi past, the claim is being made that the West German Bundeswehr has expansionist territorial goals and an offensive military strategy.

With regard to nuclear weapons, the Soviets have underscored the hostage function of any U.S. use of nuclear weapons from West German territory. The Soviet handling of the INF crisis was especially revealing of this effort. From 1978 through 1983, Soviet analysts and leaders believed that Western Europe might not actually deploy the Euromissiles. During this period, they used Soviet Euromissile capability and the Western response to that capability to try politically to separate West European from U.S. interests in NATO.

Throughout this phase of the crisis, the Soviets sought to exploit what they perceived to be a key pressure point in West-West military relations. Soviet analysts noted that the centrality of American nuclear weapons to NATO defense underscores a fundamental contradiction between Western Europe and the United States.

> The strategic calculations of the two centers of power make a strikingly different assessment of the character of a potential military conflict. In its endeavor to avoid destruction of its own territory the U.S. is oriented toward a limitation of operations to the European theater of action in the event of an armed conflict in this region, at least in its early stages. The West European military conceptions proceed from the necessity to ensure

the broad participation of the U.S. in the conflict from the very moment of its inception. This diametrical opposition of their interests is breeding a chain reaction of fears in Western Europe and disagreements with the U.S. on such questions as the stationing of American troops in Western Europe, the reliability of the U.S.'s nuclear guarantees, the role of NATO outside of its geographical bounds, the possible involvement of Western Europe in the military undertakings of American global politics.[5]

Soviet Eurostrategic capability contributes to this contradiction of interests between Western Europe and the United States. Parity in intercontinental systems tends to deter the United States from using these systems to defend Western Europe and hence raises questions within Western Europe about the reliability of U.S. nuclear guarantees.

The modernization of Soviet Eurostrategic weapons contributed to the Soviets' ability to conduct a nuclear war limited to Europe. But when the United States sought to strengthen its European theater nuclear forces, the Soviets stressed the dangers of U.S. "limited nuclear war" doctrines. They have done this in part to exacerbate West-West tensions and to impede Western military development.

To the United States, the Soviets have emphasized the impossibility of containing a Eurostrategic war without escalating to intercontinental systems. Such a claim is designed to convince the United States to cease developing theater nuclear forces for NATO's flexible response strategy. To the West Europeans, the Soviets have stressed that U.S. emphasis on a limited nuclear war shows a lack of genuine American concern for West European security. The message is that the United States is an offshore power without the same security problems as the Soviet and West European continental powers, powers that share a common home. By pressuring both the American and European components of the Alliance, the Soviets hoped to obstruct the development of Western strategic power and to slow the further development of a Eurostrategic rung on the U.S. escalation ladder.

Thus, the Soviets hoped to increase their deterrent capability, to increase their leverage on Western Europe (especially on the FRG), and to obstruct Western strategic development by maintaining their Eurostrategic superiority in the context of intercontinental strategic parity. NATO's firmness in rejecting Soviet pressures against the deployment of U.S. missiles forced the Soviets to shift their approach. Under Gorbachev, the Soviet leadership now understood that the only way to eliminate the presence of the new American missiles was to abolish completely the counterpart Soviet systems.

But the elimination of the American systems through the INF treaty still served the purpose of conveying to West European elites that U.S. and European interests were different. The Soviets have indicated to the Americans their desire for a condominium on global affairs, especially through mutual reduction of the nuclear threat to both powers. At the same time, they have expressed a desire to enhance European security significantly by finding common interests in eliminating nuclear weapons from Europe and reducing the level of conventional forces on the continent.

The pursuit of U.S.-Soviet cooperation and the enhancement of the "common home" of European security have converged in the effort to eliminate nuclear weapons from Europe. This convergence has been invigorated by the Gorbachev administration's recognition that only a plausibly serious effort at conventional arms control will permit this denuclearization campaign to proceed. Denuclearization in Europe is closely tied with the Soviets' political anticoalition strategy; it is designed to encourage the Americans to rethink their military presence in Europe and to consider the serious threat this presence poses to their own interests, narrowly considered. Furthermore, the pursuit of denuclearization in Europe will divide the European nuclear powers from the nonnuclear ones, especially France from West Germany.

Soviet analysts have underscored the centrality of the tension between the European nuclear powers and West Germany to impeding West European defense cooperation. Intra-European security cooperation has become of increasing concern to the Soviets. Such cooperation supplements the American role and could result in the creation of a two-pillar Western Alliance. The Soviets wish to avoid such an outcome at virtually any cost. Soviet leaders have underscored that Western Europe must choose between West European military cooperation and all-European detente. From the Soviet standpoint, these two options are mutually exclusive.

France and Europeanization

The Soviets perceive the evolution of French policy to be especially troubling, as France increasingly strives to strengthen the Alliance via Europeanization. Such Europeanization is to be achieved by expanding conventional military cooperation with the West Germans and nuclear military and arms control cooperation with the British. To the extent that France can encourage greater European cohesion along these lines, the effectiveness of the Soviet effort to pressure the Alliance along transatlantic lines is reduced. The Soviets would like to pursue the anticoalition strategy by exacerbating transatlantic tensions; they do not want this strategy to result in intra-European developments that would challenge their interests. But this is precisely the situation now facing the Soviet leadership and with which it is increasingly preoccupied.

The major political effect of French policy has been to stimulate the Europeanist tendency in Western Europe. Utkin has identified Europeanism as "an ideology of isolating Western Europe, of forming the West European alliance as an autonomous center in the world arena."⁶ Utkin emphasizes that Europeanism has worked at cross purposes to Atlanticism. "Europeanism as an ideology focuses on greater independence from the hegemonic power of the capitalist world in the postwar period—the United States—which results in the definite anti-American trend of Europeanism."⁷

France has played a critical role in challenging American hegemony and in encouraging the emergence of Europeanism as a practical ideology guiding the formation of a more independent West European power center.⁸ Europeanism as a trend and France's critical role in promoting it are seen by Soviet analysts as a two-edged sword. Europeanism is a positive trend to the extent that Atlanticism or American hegemony is undercut, but a negative trend to the extent that a West European power center emerges as a result. The significance of France's security policy is perceived to be intertwined with the Europeanist trend.

The withdrawal of France from the integrated military command means to the Soviets that French compliance with U.S. policy can never be assumed by Washington. The French can operate as useful interlocutors between the Soviets and the Americans or between the Soviets and other West European states. From the Soviet point of view, it is better for France to be a "perturbator" in West-West relations than a consistent supporter of U.S. positions.

The withdrawal of France from the integrated military command has also had important consequences for U.S. military policy. First, the Americans have lost the "automatic" use of French territory in a war. Because France's geographic location is critical to the conduct of a conventional campaign in Europe, the fact that it is not available to the integrated military command is seen as an important loss for the Americans. Second, the French have consistently rejected the U.S. doctrine of limited nuclear war. According to Soviet analysts, the Americans plan to fight various forms of limited nuclear war, whereas the French plan for an all-out nuclear war. The clear implication is that French nuclear forces seriously complicate U.S. plans for exercising escalation control over Western nuclear forces. Third, the French example has encouraged other Western states, such as Spain, in the direction of only partial involvement in U.S. warfighting plans. Although the French policy of independence has yielded benefits from the Soviets' point of view, the Soviets nevertheless perceive France's role in strengthening a West European power center as a largely negative one.

The sense that Atlanticism is in irrevocable decline and that the relative independence of Europe is on the rise is at the core of Soviet analyses of France in particular and of European security more generally. For Soviet analysts, Western Europe's independence can develop in one of two ways: Europe could become more independent militarily, or it could become more neutralist and embrace demilitarization. There is little doubt which of these two alternatives the Soviets favor. Equally certain is that France is contributing to the emergence of the former.

For some Soviet analysts, even the detente policy of the 1970s and early 1980s has had an ambiguous effect on the emergence of the West European power center. As an authoritative group of Soviet analysts noted, "detente is capable... of strengthening Eurocentric tendencies in the military policies of the West European powers, both within the framework of NATO and outside it. The Europeanization of NATO, i.e., the increase in the relative importance within it of its European participants, may strengthen."[9] The dark side of France's detente policy has also been noted: "The French government believed that the detente process had to parallel a strengthening of each West European country and all of Western Europe as a whole."[10]

Thus, the significance of French security policy to Soviet analysts is due largely to its contribution to the Europeanist tendency. Although France's example of independence has been important in stimulating the West Europeans' quest for greater independence from the Americans, France's contribution to security interdependence within the West European power center worries the Soviets more. Soviet analysts have noted a number of ways in which the evolution of French policy will decisively affect the emergence of security cooperation within Western Europe. First, to the extent that France develops its cooperation with Britain, a true European nuclear force (a Soviet term) could emerge.[11] Second, as the Franco-German security relationship becomes stronger, the Germans are more likely to acquire nuclear weapons, indirectly or directly.[12] Third, to the extent that the major West European powers—perhaps through the Western European Union—cooperate in joint armaments production, the possibility of greater cooperation in conventional force deployments is enhanced. Weapons standardization enhances the probability of conventional military cooperation.

At least some Soviet analysts have posed the possibility that Atlanticism and Europeanism might not be polar opposites. Rather, Europeanism might provide the Americans with more room to maneuver in dealing with contingencies outside Europe. Actions taken by the Mitterrand administration to promote European cooperation and to reduce the anti-American thrust associated with traditional Gaullism are a manifestation of something approaching a worst-case scenario for the Soviets.[13]

The Role of Great Britain

The concern with the emergence of an enhanced European nuclear capability is reflected not only in Soviet assessments of the French but of the British as well. Great Britain's "special nuclear relationship" with the United States encourages the British to be key supporters of the Americans in the European nuclear debate. Soviet analysts underscore that U.S.-UK nuclear sharing arrangements allow the British to be the best informed of all America's European allies concerning U.S. nuclear policy. Also, the British have been able "to influence to some extent the United States' course in elaborating NATO's nuclear strategy, to receive information first-hand about a broad range of problems, and to participate in solving them."[14] Being kept in the nuclear ballgame through U.S. assistance, the British are generally the most pronuclear and pro-American of the U.S. allies within NATO. The British are especially strong supporters of maintaining the U.S. nuclear guarantee to Europe. Finally, by remaining a nuclear power, Great Britain not only supports U.S. nuclear policy but also rejects Soviet options for reducing tensions associated with the nuclear competition.

Not only has the British emphasis on nuclear deterrence been troubling to the Soviets, so too has been its growing interest in the Europeanization of NATO. Soviet analysts argue that the British have a twofold approach to the Europeanization of the Alliance. On the one hand, the British try to act as the interlocutor between Europe and the United States and to use their position in NATO to generate greater influence within Europe.[15] On the other hand, the British seek to encourage the development of a "European pillar" within NATO. They helped to create the Eurogroup and the Independent European Program Group (IEPG) to facilitate this process. They have developed extensive ties with other European states in joint armaments production in order to facilitate Europeanization as well.

In the Soviet view, the British play the preeminent European role in NATO, as illustrated in their high degree of representation within the command structure and in various NATO committees. The British approach to Europeanization is to draw on their strengths in NATO in order to assert their influence within Europe. Kolosov has argued that all this high-level influence, combined with "the significant 'presence' of lower rank British officers and specialists in NATO's headquarters, commands, and various committees, is called upon to protect the specific interests of England's government and military leadership and not to permit a diminution of its influence within the bloc's military organization."[16]

The British have sought to Europeanize the Alliance through the formation of such institutions as the Eurogroup. According to Madzoevskii and Khesin, "A definite 'Europeanization' of its [Britain's] nonnuclear component is taking place under the auspices of the Eurogroup's programming, which is uniting the majority of the European members of NATO (without the participation of the U.S. and Canada)."[17] The NATO Eurogroup is perceived by Kolosov to play an "extremely" vital role in the coordination of collaboration efforts among the West Europeans and has "contributed to the increase of their contribution to the build-up of NATO's military might and to the development among them of military-policy, military, and military-industrial ties."[18] The Eurogroup, however, has acted only in certain spheres to promote collaboration, namely in military-technical rather than military-policy tasks.[19]

The linkage between the Eurogroup, Europeanization, and the significance of European—including British—conventional forces to greater inter-European cooperation has been clearly articulated by Kolosov.

> In the military arena, England, the FRG, Italy and the other West European members of NATO provide the main portion of general purpose forces that are focused on executing the bloc's tasks.... There is a growing potential for developing more durable and long-term ties among the national ministries of defense, the armed forces in the most diverse fields—from their material-technical support to the conduct of joint maneuvers, and the elaboration of, to all intents and purposes, unified operational tactical plans for conducting a war.[20]

The British have not been receptive to Europeanization approaches outside the NATO context. From the Soviet point of view, the reason Britain has rejected Europeanization outside NATO has been the priority the British place on NATO's coalition military strategy. By and large the British perceive their forces to be most effective when operating jointly with NATO's forces. The British have drawn the Americans into a tighter nuclear guarantee by allowing deployment of U.S. nuclear weapons on British soil and promoting nuclear interdependence with the United States. The British have, however, retained the ultimate capability to use their nuclear weapons to serve national interests.

The Soviets expect the British orientation toward Europeanization to emphasize the nuclear component of European defense. Many Soviet analysts of post-Reykjavik developments were struck by the extent to which conservative European governments worked together to ensure that the Americans stayed on the nuclear path. For example, the Soviet press was quick to report on British Prime Minister Margaret Thatcher's visit to Paris in November 1986, when she sought to enhance her ability to influence the Americans on nuclear disarmament issues through a unified Franco-British stance.[21]

Nonetheless, Soviet analysts are well aware of the tensions confronting intra-European cooperation as West European leaders seek to Europeanize the Alliance. This was especially striking in one analyst's comment about the effort to revitalize the Western European Union.

> At the creation of the WEU, the leadership of the main West European states hoped to use it as a means for the development of cooperation

between the member states and for the increase in their role for determining the policy of NATO. According to the [1954] Paris agreement, this Union was to facilitate the unity and gradual integration of the West European states in the military-political sphere, the setting up of joint weapons production and their standardization. However, in reality, according to foreign observers, it turned into a merely consultative organ whose decisions do not have an obligatory character for the member states. Furthermore, to a significant degree the effectiveness of the WEU activity is lowered due to a constant collision of interests of Great Britain, France and the FRG waging a fierce struggle for leadership in Western Europe.[22]

The East European Factor

The Soviet effort to limit Atlantic cohesion and to abort a Europeanization that emphasizes military cooperation is reduced in its effectiveness by Soviet domination of Eastern Europe. Since the end of World War II and the Red Army's occupation of the region, Eastern Europe has held a special place in Soviet foreign policy. Given Stalin's desire to turn "socialism in one country" into "socialism in one region," Eastern Europe became, and has remained, more than a sphere of Soviet influence. It became a sphere of control, designed to demonstrate the universal applicability of Soviet-style socialism.

If the Soviets had opted instead for a ring of neutral, semi-independent states in Eastern Europe, several benefits might well have resulted. First, the East European countries would likely be more stable and their peoples less anti-Soviet than is currently the case. In addition, neutrality in this region would have better served Soviet interests and influence among the West European countries.

Yet the concept of having to control Eastern Europe persists even today. From Stalin to Gorbachev, Soviet leaders have perceived control to be necessary in order to have influence. Admittedly, Gorbachev's language and tone differ from those of his predecessors,[23] but the basic Soviet concepts remain the same. This is not to say, however, that absolutely nothing has changed. There has been an evolution in Moscow's policies toward Eastern Europe over the past several years. Nevertheless, what remains fundamentally unchanged is the Soviet perception of its interests in the region.

With respect to the Soviet Union's policy toward Western Europe, relations with Eastern Europe pose a problem and a liability. Fear of losing control over Eastern Europe is the main reason the Soviet leadership has pursued a policy of isolation and intimidation of Western Europe rather than of seduction. The question for the present and the future is whether Gorbachev and his advisors will prove willing to alter the traditional Soviet approach toward Europe.

The Gorbachev Leadership and the Public Diplomacy Dimension

Although overall Soviet objectives toward Western Europe have remained the same for several years, the Gorbachev leadership has injected a new dynamism in Soviet foreign policy whereby the implementation of these objectives is being pursued differently and more vigorously than in the immediate past. The Soviet leadership under Gorbachev has sought a "breathing space" in East-West relations to aid their domestic reform efforts. In order to obtain a breathing space, however, the general secretary believes it necessary to convince the West that he is seeking fundamental reform in the Soviet Union. In other words, to ensure the breathing space necessary for economic modernization, it is helpful to convince the West that real change is occurring in the Soviet Union. Gorbachev has done this through public diplomacy and through summit meetings with Western leaders.

It is in the area of Soviet public diplomacy that the greatest change has occurred in Soviet foreign policy, particularly with respect to Gorbachev's objective capabilities. The general secretary is the first Soviet leader to make effective use of the electronic media; he instinctively understands how to use the media to his political advantage. This approach holds true in both his domestic and foreign policy efforts.

But the Gorbachev phenomenon signifies more than the ascension of a media personality to the Soviet leadership. The Soviet system has seen the development of an increasingly sophisticated foreign policy apparatus. A new foreign policy elite has been emerging over the past 15 years that is better trained and better informed than its predecessors. The previous dominance of Andrei Gromyko in Soviet foreign policy and the general aging of the Soviet elite impeded the emergence of this new generation. Gorbachev's ascension to power has enabled the new elite to rise to positions of power in the foreign policy system, a process that will continue over the next few years.

Gorbachev has largely subsumed the thinking of these new policymakers. In effect, the new elite has provided the language and ideas for the general secretary, who has used them to promote his image and the place of the Soviet Union in the global system. Furthermore, Gorbachev and Foreign Minister Eduard Shevardnadze are aware of the benefits of a flexible leadership style and good public relations at home and abroad, unlike Shevardnadze's predecessor, the increasingly inflexible and dour Gromyko.

Gorbachev has also altered the actual structure of foreign policymaking. He has ended the dominance of the Foreign Ministry over foreign policy. Shaking up the foreign policy elite and opening up the system to new faces have led to a ferment in foreign policy activity. These actions have also allowed the Soviets to rethink some old positions and to move forward more flexibly. For example, it is clear in retrospect that the inflexibility of the Soviet negotiating position on the INF talks was due largely to Gromyko. Gorbachev proved willing to rethink the Soviet position and to accept the legitimacy of Western demands for a serious reduction in Soviet INF systems.

The general secretary has also tried to promote an image of flexibility in pursuing a policy of summitry with Western leaders. He has conveyed to these leaders his seriousness about domestic change, as well as the priority he places on domestic modernization. He has shown also that he understands the importance of projecting to the West an image of flexibility, reciprocity, and candor—the foreign policy equivalent of *glasnost'*.

In short, a new dynamism has been evident in Soviet foreign policy, which has been especially apparent in Soviet public diplomacy toward the Alliance. An important component of Soviet public diplomacy in Western Europe under Gorbachev has been a new realism. Rather than focusing merely on political forces more favorable to Soviet interpretations of Western interests (such as the peace movement), Soviet leaders are beginning to recognize the staying power of the more conservative elements in Western Europe. This shift is due primarily to the recent conservative party victories in both the United Kingdom and West Germany. Because it is now clear that in many of the key countries conservative governments will be in power into the 1990s, the Soviets recognize the need to reach out to a broader elite in Western Europe. For example, in West Germany, whereas in the past Soviet officials would court primarily Social Democratic Party (SPD) officials, now such a conservative politician as Franz Joseph Strauss has been welcomed in Moscow. In addition, West European leaders and political forces with close connections to Washington are treated as potential conduits to Washington. For example, Thatcher clearly is seen by Gorbachev to be a useful interlocutor in this light.

The Soviet embassies in Western Europe are actively engaged in courting the Western press with much more candor and vigor than in the past. The Soviets seem to understand the critical role that Western journalists play as gatekeepers for public perceptions of foreign policy. Soviet embassies in Western Europe have encouraged the more conservative West European newspapers to engage more actively in reporting Soviet events directly.

When interacting with West European elites, the Soviets have relied increasingly on a new type of influence agent, namely the *institutniki*— prominent academic specialists—and other members of the intelligentsia fluent in European languages. Increasingly, the Soviet intelligentsia who travel to Western Europe are well trained in the professional language of the Western elite. Compared to the past, they are more capable of arguing the Soviet case in Western terms, using Western (especially American) data.

Another important dimension of the exercise of public diplomacy under Gorbachev has been the general attempt to bring words more into line with deeds. Rather than relying on wishful thinking, clichés, or bombastic propaganda alone, the Soviets have begun to recognize that changes in Soviet actions are required to effect changes in West European perceptions. This is especially true in both the foreign economic and arms control arenas. With regard to foreign economic activity, the Soviet leadership is in the throes of implementing major changes in their foreign trade organizations. Associated with these changes are efforts to increase the involvement of foreign firms in joint ventures with Soviet firms. Soviet commentators have highlighted the significance of such changes in making involvement in the Soviet economy more attractive to West European firms. Especially active in this selling campaign have been prominent Soviet economists and other members of the intelligentsia who are likely to have much greater credibility than Soviet political officials alone.

In the arms control arena the Soviets have sought to carry out actions the West would recognize as significant; that is, actions that would indicate that the Soviets are serious about lowering the level of confrontation in East-West relations. The acceptance of the single zero option, the promotion of the double zero option, and then the acceptance of Western insistence on an on-site verification regime have significantly enhanced the perceptions that the Soviets are seeking genuine reconciliation in East-West relations. Generating change in foreign policy actions provides significant grist for the public diplomacy mill.

The Soviets have developed a number of important themes in their public diplomacy toward the Alliance during the Gorbachev period. Above all, they have promoted the idea that there is now a "new" look in Soviet policy. Soviet officials are now more realistic and willing to accept the realities of modern society as they understand it. The Soviets claim to accept the legitimacy of a number of Western institutions that were once anathema to them, such as NATO, the EEC, and the WEU.

While the Soviets might not like any of these institutions, they do exist, and the Soviets have finally recognized that they can no longer ignore them. For example, with respect to NATO, high-ranking Soviet officials (including Gorbachev himself) have frequently stated during the past three years that they are not trying to drive wedges in the Alliance and that they do accept the legitimacy of some form of American presence in Western Europe.

Furthermore, the Soviets have underscored to West Europeans that the nature of Soviet society is changing. The current image is of a society that had been stultified by the old but that has now been unleashed by new thinking. During the past 30 years the West European public has perceived the vitality of Soviet society to be in steady decline; by the end of the Brezhnev period only a small minority of West Europeans viewed Soviet society in favorable terms. The Soviets are explicitly trying to reverse this trend by using the "new thinking" and *glasnost'* campaigns to highlight innovation in Soviet society.

In addition, Soviet spokespersons use Gorbachev's favorable image to promote the belief that fundamental reconciliation in East-West relations is now possible. Gorbachev's book, entitled *Perestroika: New Thinking for Our Country and the World*, published in 1987, has been distributed widely in Western Europe and the United States and is perceived by many West Europeans to promise the possibility of a serious reduction in East-West tensions for the first time in a long while. Underscoring Gorbachev's quality as a great statesman is a critical component of the Soviet public diplomacy campaign in Western Europe today.

Another major theme is that of Western Europe and the Soviet Union sharing a common European home. Although the concept did not originate with the Gorbachev administration, it has now become a hallmark theme of its current effort in Western Europe. The goal is to convince the West Europeans that they share many interests in common with the Soviet Union and that if a significant effort is made to overcome old stereotypes, progress toward genuine interdependence is possible.

Associated with this theme is the growing recognition of the legitimacy of West European economic integration, although hostility to political and military integration remains strong. In fact, the Soviets counterpose all-European security cooperation to West European military integration. In this connection, Soviet writings underscore the threat that any enhancement of West European security cooperation poses to genuine detente. For example, the Soviets have focused growing attention on the Franco-German security partnership and have begun to target ways of undermining it.

In the wake of the INF agreement, the Soviets are also stepping up their campaign for nuclear disarmament in Europe. From the Soviet perspective, nuclear weapons remain a critical threat to the homeland; the Soviet leadership therefore seeks to reduce that threat through various denuclearization proposals. In addition, it perceives nuclear weapons to be the linchpin of America's involvement in its military alliances, especially in NATO. By promoting the desirability of further progress toward denuclearization in the European theater, the Soviets might ultimately succeed in undermining the American military presence in Western Europe.

To bolster the chances for denuclearization, the Soviets are now underscoring the need to make real progress in conventional arms control. They have admitted the existence of Soviet conventional superiority in some areas in the European theater and have demonstrated their resolve to back up their words with deeds, as evidenced in Gorbachev's conventional force reduction announcement to the United Nations on December 7, 1988. The 1987 Warsaw Pact declaration on military doctrine and conventional arms control has been prominently promoted in Soviet interactions with West Europeans precisely to underscore the seriousness of Soviet interest in conventional disarmament. This effort is used, in turn, to strengthen the denuclearization campaign in Western Europe.

An important component of the conventional arms control process has been the desire to legitimize the ideas of the European left on security issues. During the INF deployment struggle of the early 1980s, the left opposed not only nuclear but conventional modernization as well. Their opposition to conventional modernization was grounded in the concept that defense should be genuinely defensive, not oriented toward "provocative" offensive operations. Several Soviet spokespersons now accept the legitimacy of these ideas under the rubric of "reasonable sufficiency." There is an explicit connection between the Soviet argument about the need to shift Soviet doctrine and forces to a posture based on reasonable sufficiency and the West European left's promotion of the idea of nonprovocative defense.

Above all, Soviet commentators are underscoring that there is a growing and more realistic possibility of East-West reconciliation if Soviet initiatives are met with correspondingly serious Western proposals. Simply put, the message is: "We are now reasonable; work with us and we can put the old confrontations behind us and live in a more peaceful and interdependent world."

In short, Soviet public diplomacy is significantly aided by the new dynamism of Soviet foreign policy. It is aided as well by the new realism in Soviet assessments of developments in the West and by the promotion of the Soviet Union's image as a society in crisis, and one that thereby no longer poses a serious threat to the West.

The Military Strategy Dimension

The Soviet approach to the Alliance attempts to combine a political-military and military-technical strategy. The political-military strategy revolves around an anticoalition approach. It also underscores the salience of the political aspects of military competition with the West and the need to leverage the Alliance better in times of crisis and war. The military-technical policy emphasizes the need to be able to prevail if war comes in Europe through the use of conventional forces and the threat of immediate escalation in the event of nuclear use by NATO.

Marshal Akhromeev's formulation of coalitional strategy argues that the lessons of World War II are important and relevant to Soviet policy today. "The main lesson of World War II, namely that war must be combated before it has begun, therefore assumes special topicality today. Historical experience indicates that joint, concerted, and vigorous action on the part of all peace-loving forces against the aggressive actions of imperialism is necessary in order to defend peace."[24]

Central to Soviet thinking is the need to prevail in the initial period of any future war. The salience of political factors to such an effort is also taken into account. Among the critical political factors affecting success in the initial period of war are these: the ability to mobilize forces, the ability to create a general transition from a crisis to armed conflict, and the ability to deal with the issue of nuclear escalation.

With regard to mobilization, the Alliance members will undoubtedly find themselves in conflict over whether to mobilize forces, over which forces to mobilize, and over whether the Soviet Union is really prepared to go to war. During the crisis period, the mobilization decision will become tantamount to shifting to war. Colonel-General M. A. Gareev, for example, has argued that there is an almost irreversible nature inherent in the mobilization process and that this process will be deeply affected by the political environment. "If a war generally is politics through and through, on the eve and at the start of a war its political aspects are even more prevalent."[25]

The general transition from peacetime to wartime will be influenced profoundly by political conflict within NATO. Differing national interests may come to the fore, which the Soviets would seek to exploit. A significant disinformation campaign will be unleashed in the political arena to compound NATO's difficulties. Soviet military analysts refer to Nazi efforts as models for success in the transition phase from crisis to war. According to Matsulenko:

> The Nazi leadership carried out a large range of measures involving virtually all the bodies of state and military administration, all means of mass information and the diplomatic corps. Here the main goals of the political actions were to conceal the very fact of the aggression being prepared and to prevent the nation which was to be attacked from promptly discovering the danger threatening it. The surprise and deception were aimed at concealing the very measures related to organizing the aggression, and in particular the strategic deployment of the armed forces, the axes of the main thrusts and the time of attack. The most limited number of persons was involved in working out the operational-strategic planning documents and measures were taken to mislead the enemy about the place, time and methods of actions.[26]

After hostilities have begun, the Soviets will seek to prevail at the conventional level. Nonetheless, the campaign would be fought under the constant threat of nuclear use. The political dimensions of the nuclear decision are significant in Soviet thinking. Despite Soviet claims of an "all or nothing" strategy for nuclear deterrence, Soviet analysts have contemplated the use of limited nuclear strikes for political purposes.[27] Lectures at the

Soviet General Staff Academy in the mid-1970s taught students that "political actions may affect the selection of the TVD [theater of military action] for action, the selection of the countries to be hit by nuclear strikes, or the nations not to be attacked by nuclear weapons."[28]

From the standpoint of an anticoalition political-military strategy, the Soviets might conduct their military campaign in such a manner as to put primary pressure on those states with forward-deployed forces in West Germany. They would concentrate on those countries that they perceive to be the weak links in NATO (the Netherlands, Belgium, and Canada) and would encourage these states to withdraw from German soil. The Soviets might define the threat as American "militarism" and West German "revanchism," the threats against which they *must* act. Soviet leaders would convey through diplomatic and propaganda channels that they have no hostile intentions against other European states, especially European states with forward-deployed forces. If the Soviets could get one state to withdraw these forces, they would hope to set off a chain reaction of withdrawal. They would hope especially, by means of such a chain reaction, to pressure the British to withdraw. Given the U.S.-British special relationship, a British withdrawal of forward-deployed forces might be seen as able to affect U.S. attitudes and policies significantly. Moreover, the Soviets could more credibly offer nonintervention pledges to the British than to any other major European Alliance power.

The key Soviet objective would be to isolate the FRG as much as possible. If some forward-deployed forces could be pressured diplomatically to withdraw, an isolation process would be started. The Soviets would hope to freeze West German mobilization and West German willingness to support reinforcement efforts as long as possible. The Soviets would convey to the German chancellor that significant mobilization would be considered an act of war, thereby raising the risk of mobilization from the outset.

The Soviets would also seek to pressure France to follow its traditional strategy of protecting French territory. The Soviet leaders would clearly encourage the French to keep the Force d'Action Rapide in French territory and to remove their forces from Germany in exchange for a nonaggression pledge. A key question for Soviet leaders would be whether to attack French nuclear forces. Would frontal attacks or restraint be more effective in encouraging French acquiescence in West German neutrality?

The Soviets might also prosecute (pursue and/or attack) British and French nuclear ballistic submarines (SSBNs). They would do so to reduce the nuclear threat against the Soviet Union. Successful prosecution would also increase the willingness of the two countries to withdraw from the conflict and pledge neutrality and support for the demilitarization of West Germany. Success in the destruction of French and British SSBNs might also send a useful signal to the United States, encouraging U.S. leaders to husband SSBN assets rather than use them in limited nuclear strikes.

As the Soviets begin military operations against NATO forces in West Germany, they would hope to encourage key Alliance states to revert to strictly national strategies. From the outset, they might declare limited objectives against NATO forces in Germany and, if successful in capturing specific concentrations of national forces, might exchange prisoners for pledges of neutrality by the specific country. Throughout the German campaign, the Soviets would seek to influence U.S. policy and actions through successes on the battlefield. The Soviets would try to destroy or isolate U.S. forces and then perhaps seek U.S. pledges for West German demilitarization as well.

Nonetheless, conventional operations in the European theater would be only quasi-conventional in character. The Soviets would seek to eliminate as much of NATO's nuclear potential in Europe as possible. Their key goal during conventional operations would be to degrade Western military forces in Europe, especially nuclear forces. The Soviets have become increasingly sensitive to the possibility and even the desirability of withholding nuclear strikes as long as possible or, in other words, of prolonging conventional operations in a war with the West as long as it is feasible to do so. Thus, a key military objective of Soviet nuclear weapons in Europe is to make a conventional warfighting option more viable.

The Soviets might contemplate the use of various types of nuclear weapons on the European battlefield to support limited warfighting objectives. The Soviets might use their short-range battlefield nuclear weapons to attack NATO ground forces, especially NATO's short-range battlefield weapons. The basic political objective would be to increase the level of damage on West German territory and thereby to encourage West Germany to pressure NATO for an early end to nuclear hostilities.

The Soviets might go further, using their longer range theater nuclear force systems (notably their aircraft, such as the Fencer and Backfire) to attack high-value NATO military targets in Europe. Among key NATO targets would be airfields, ports, and tactical missile and warhead sites. In addition to the obvious military objectives of such an attack, a number of political objectives are served as well. The Soviets would be pressuring NATO for a termination of hostilities prior to conducting significant strategic attacks directly against the territory of the European nuclear powers, Great Britain and France. A subordinate goal might be to try to neutralize France and Great Britain in order to undercut their will or need to use their strategic weapons against Soviet territory.

By such actions, the Soviets would seek to influence U.S. escalation decisions. As European theater nuclear strikes occurred, the Soviets would seek to assure the United States that the war could be confined to NATO Europe. Throughout, the Soviets would seek to encourage the United States to desist from intercontinental nuclear exchanges.

The next chapter addresses some of the military-technical factors briefly raised here. The remainder of this book then concentrates on the variety of political-military factors that comprise the current Soviet approach to the Western Alliance. An appreciation for both these components—the military-technical and the political-military elements—is critical to understanding Soviet policy toward the West.

Notes

1 For earlier treatments by the author of the general Soviet approach to the Alliance, see the following: Erik P. Hoffmann and Robbin F. Laird, *"The Scientific-Technological Revolution" and Soviet Foreign Policy* (Elmsford, NY: Pergamon Press, 1982); and Laird, *The Soviet Union, the West and Nuclear Arms* (New York: New York University Press, and London: Wheatsheaf Press, 1986).
2 See, for example, Aleksandr Bovin in *FBIS*, September 25, 1985, p. G2.
3 *FBIS*, November 25, 1986, p. AA3.
4 See the exchange between Viktor Tsoppi and Aleksandr Antsiferov in *FBIS*, November 24, 1986, p. AA7.
5 V. N. Shenaev, ed., *Western Europe Today* (Moscow: Progress Publishers, 1980), p. 306.
6 A. I. Utkin, *Doktriny atlantizma i evropeiskaya integratsiya [Atlanticism's Doctrines and European Integration]* (Moscow: Nauka, 1979), p. 73.
7 Ibid.
8 V. P. Lukin, *"Tsentry sily": kontseptsii i real'nost' ["Power Centers": Concepts and Reality]* (Moscow: Mezhdunarodnye otnosheniya, 1983), pp. 66–67.
9 V. N. Shenaev, D. E. Mel'nikov, and L. Maier, eds., *Zapadnaya Evropa: ekonomika, politika, klassovaya bor'ba [Western Europe: Economics, Politics, Class Struggle]* (Moscow: Mysl', 1979), p. 261.
10 T. I. Sulitskaya, *Kitai i Frantsiya, 1949–1981 [China and France, 1949–1981]* (Moscow: Nauka, 1983), p. 139.
11 For the fullest statement of the European nuclear force concept, see V. F. Davydov, "Discussion of the European Nuclear Forces," *SShA*, 3 (1976), pp. 28–38.
12 See, for example, V. Mikhnovich, "Calculations and Miscalculations," *KZ*, June 14, 1984, p. 3.
13 See, for example, Capt. V. Kuzar', "NATO—An Alliance in the Name of Aggression," *KZ*, April 8, 1984, p. 3.
14 G. V. Kolosov, *Voenno-politicheskii kurs Anglii v Evrope [England's Military-Political Course in Europe]* (Moscow: Nauka, 1984), p. 65.
15 Ibid., p. 10.
16 Ibid., p. 106.
17 S. P. Madzoevskii and E. S. Khesin, "Great Britain in Today's World," *MEMO*, 8 (1980), p. 56.
18 Kolosov, *Voenno-politicheskii kurs*, p. 178.
19 Ibid., p. 184.

20 G. V. Kolosov, "The Military-Political Aspects of the West European Integration Process," in N. S. Kishilov, ed., *Zapadno-evropeiskaya integratsiya: politicheskie aspekty [West European Integration: Political Aspects]* (Moscow: Nauka, 1985), p. 246.
21 Yu. Kovalenko, "M. Thatcher's Two Trips," *Izvestiya*, November 23, 1986, p. 4.
22 Col. I. Vladimirov, "Basic Tendencies in the Military Integration of the West European Countries," *ZVO*, 3 (1982), p. 13.
23 Compare, for example, Brezhnev's statement with Gorbachev's, *Pravda*, November 13, 1968, and April 11, 1987.
24 Marshal S. F. Akhromeev in *FBIS*, May 13, 1986, p. 8.
25 Col. Gen. M. A. Gareev, *M. V. Frunze—voennyi teoretik [M. V. Frunze—Military Theoretician]* (Moscow: Voenizdat, 1985), p. 242.
26 V. Matsulenko, "Certain Conclusions from the Experience of the Initial Period of the Great Patriotic War," *VIZ*, 3 (1984), p. 38.
27 See the author's book coauthored with Dale R. Herspring, *The Soviet Union and Strategic Arms* (Boulder, CO: Westview Press, 1984).
28 "Principles of Strategic Action of the Armed Forces," *The Voroshilov Lectures: Materials from the Soviet General Staff Academy* (Washington, DC: National Defense University Press, 1989).

2
Soviet Views and Policies toward Theater War in Europe[1]

Phillip A. Petersen
Office of the Secretary of Defense

Notra Trulock III
National Defense University

Soviet military theory and practice have focused on the conduct of theater warfare for most of the postwar period. Although the Soviets have not neglected other modes of conflict, even during the focus on global nuclear war in the early 1960s, they continued to believe that their main strategic objectives were on the Eurasian landmass. The achievement of strategic nuclear parity with the United States and the impact of this on the U.S. policy of extended deterrence, however, permitted the Soviets to intensify their planning for the conduct of theater war around the periphery of the Soviet Union.

Because it is in Europe that the superpowers and the most powerful military coalitions confront each other directly, Europe has retained its centrality within the Soviet framework for the planning and execution of theater operations. That framework considers the strategic operation in a continental theater of strategic military action (TSMA—*teatre voennykh deistvii*, or TVD)[2] to be one of the basic forms of strategic action. Based on their net assessment of the military geographic situation of the opposing coalitions in Europe, Soviet military theorists have divided Europe into northwestern, western, and southwestern TSMAs. For obvious reasons, the Soviets consider the western TSMA to be the probable main TSMA in a war between the two coalitions since within its boundaries are located the most powerful military groupings in NATO.

This chapter discusses how Soviet military theory and practice for theater war in Europe have developed over the last two decades. Following an overview of the central emphasis on Europe and the continuity of wartime strategic objectives in Soviet military planning, the chapter traces the evolution of Soviet concepts and methods of achieving these objectives in the event of such a war.

As Soviet military planning emerged from the "one-variant war" phase of the early 1960s, Soviet military planners concluded that near-term technologies afforded an opportunity—given the right conditions and skillful applications—to achieve their strategic objectives without resort to nuclear weapons. A central dilemma for Soviet military planners, however, has been the overriding necessity to restrain the threat of the enemy's use of nuclear weapons. Soviet approaches to this objective have been multiple and varied. The denuclearization of Europe has been a key Soviet military-political objective since the 1950s and remains so today. This effort notwithstanding, the Soviet military has continued to develop operational concepts and combat capabilities intended to inhibit NATO's resort to nuclear weapons and to reduce the impact of any nuclear weapons that are used. Obviously, the Soviets have also been forced to continue to consider the requirements associated with a transition to the use of nuclear weapons. The manner in which Soviet military planners have sought to overcome the dilemma of fighting a conventional war under the threat of the enemy's use of nuclear weapons represents the central focus of this chapter.

The final section of the chapter examines the Soviet forecast of an imminent qualitative leap in military affairs. This forecast encompasses the implications of emerging military technologies and ongoing doctrinal developments in both NATO and the Warsaw Pact. Although the Soviets believe emerging military technologies have already created a transitional stage, they also recognize that the limited availability of these new weapons may stretch this stage well into the 1990s. Nevertheless, the Soviets are intent not only to find early solutions to the most effective exploitation of these new technologies, but also to minimize the impact of NATO's acquisition of these weapons on future Soviet strategic operations in a continental TSMA.

Soviet Perceptions of Military Power and Deterrence in Europe

The Soviet Union has no more incentive to achieve its political objectives in Europe through war than does the West. These objectives are pursued, instead, through a combination of diplomatic, arms control, economic, and perception management instruments. The Soviets have, however, never lost sight of the value of tangible military power in achieving their foreign policy objectives, although they have recently sought a better balance between military power and other instruments of national security policy. A 1972 Soviet assessment of the benefits of military power, for example, concluded that "changing the correlation of military forces . . . is having a sobering effect on the extremist circles of the imperialist states and is creating favorable conditions for the realization of Soviet foreign policy objectives in the world arena."[3] These benefits were derived not just from the perceived impact of the achievement of strategic parity with the United States, but also from a concerted effort throughout the ensuing decade to discredit NATO's strategy and force-development programs.

Prior to the advent of strategic parity in the late 1960s, the Soviets believed they were forced to rely on a "hostage Europe" approach in order to obtain a sobering effect on Western decision-makers.[4] Although some skepticism has been expressed in the West regarding this interpretation of Soviet military policy, the Soviets seem to have adopted this approach and, in fact, have described it in such terms. An *Izvestiya* reprint of an interview given by then Premier Nikita Khrushchev to the *New York Times* in September 1961 makes this strategy clear. "Khrushchev believes absolutely that when it comes to a showdown, Britain, France and Italy would refuse to join the United States in a war over Berlin for fear of their absolute destruction. Quite blandly he asserts that these countries are, figuratively speaking, hostages to the USSR and a guarantee against war."[5]

The deterrent value of the Soviet intermediate-range nuclear missile force (INF) was evidently intended by Khrushchev as a stopgap measure to hold Europe hostage until Soviet deployments of intercontinental ballistic missiles (ICBMs) were sufficient for the threat of their use to restrain the United States from using nuclear weapons in the defense of Europe. The deterrent value of the rapidly growing nuclear missile forces and the Soviet ground forces was made even clearer by Marshal P. A. Rotmistrov in an unguarded moment at the June 4, 1965, Finnish Army Day reception in Moscow. Rotmistrov is reported to have argued along the following lines:

> The Soviet Army is a continental power. It must maintain control of the European continent. Now, the U.S. has tremendous nuclear capabilities adequate to destroy much of the Soviet Union. The latter's is somewhat more modest, but still sufficient to damage the U.S. to the degree that one must take it into consideration. As time passes, the threat to the U.S. grows greater, and eventually, deterrence will become as binding on the U.S. as on the Soviet Union. Thus, the Soviets, with a valid counterstrike capability, will continue to maintain their ability to overrun Europe in 60–90 days either in a nuclear or a nonnuclear situation. Thus, Europe will remain hostage to the Soviet Army.[6]

Until this time, the lingering effect of the "massive retaliation" policy that was declared during the Eisenhower administration continued to orient Soviet planning scenarios toward surprise massive nuclear strikes by the United States against the nascent Soviet intercontinental nuclear forces. The development of the Soviet Union's "somewhat more modest" nuclear forces had, however, forced the United States and NATO to undertake the transformation of their own strategy. An internal 1966 Soviet military assessment of this transformation concluded that "the development of thermonuclear weapons and the appearance of intercontinental means for their delivery made a general nuclear war such a dangerous means of policy that a reconsideration of the entire NATO strategy was required."[7]

After forcing the West to abandon the strategy of massive retaliation, Soviet military planners turned their attention to the task of discrediting the new U.S.-NATO strategy of flexible response. This challenge was to be met through the development of a balanced conventional and nuclear force posture. The intent was to develop both an observable capability to defeat NATO conventionally and, at the same time, the ability to restrain NATO from the use of nuclear forces by deploying Soviet nuclear forces sufficient to absorb the initial strike and still respond decisively. Such a capability could effectively undercut the premise of flexible response and narrow the range of rational choice open to NATO decision-makers in the event of a deepening crisis or the outbreak of hostilities. Efforts to defeat NATO conventionally simply used and expanded existing capabilities and concepts, while those directed at preventing nuclear use by NATO involved a combination of force-development and perception-management activities.

A specific objective of Soviet policy has been to dissuade NATO from even considering nuclear responses to Soviet conventional operations in the event of a European war. This was to be achieved, in part through the deployment of large, successively more accurate and survivable theater nuclear forces. Soviet military planners clearly believed that just the threat of using these forces in a large-scale, indiscriminate fashion in response to even a limited NATO nuclear strike could have a decisive impact on NATO decision-making. For example, in a lecture to the students of the Voroshilov General Staff Academy, then Academy Commandant Army General I. E. Shavrov argued that "the possibility cannot be excluded that the danger of massive strikes by all nuclear weapons in retaliation for any attempt at the use of nuclear weapons, be it of a limited nature, may force the imperialist countries to give up the continuation of the combat actions."[8]

General Shavrov's reference to "all nuclear weapons" is indicative of the Soviet attempt to persuade NATO that the Soviet Union would not accept any form of limitation on the employment of its own nuclear forces and, instead, would readily resort to the use of not just theater forces but of intercontinental systems as well. Such a perception on the part of U.S. and NATO decision-makers, in the Soviet view, could not fail to diminish their willingness to consider nuclear responses to a Soviet conventional attack. Furthermore, if NATO should make limited use of nuclear weapons, the Soviet threat of massive retaliation might be sufficient to convince some NATO members to withdraw from the Alliance and the war. Not surprisingly, the Soviets have invested enormous efforts to ensure this message is received by Western decision-makers, elites, and Soviet watchers.

Such deliberate dissimulations are consistently transmitted through public and private elaborations of Soviet views on the logic of nuclear war by senior Soviet military authorities.[9] Nearly all such spokespersons agree with Marshal N. V. Ogarkov's repeated assertion that "any limited use of nuclear weapons will inevitably lead to the immediate use of the sides' entire nuclear arsenals. Such is the harsh logic of war."[10] This theme is consistently reinforced by Soviet civilian national security authorities with good Western media access.

The Soviets may also seek to convey this message through their training practices. For example, the use of exercises to convey disinformation to the enemy regarding Soviet intentions and plans has a long foundation in Soviet military theory, and operational disinformation was effectively and repeatedly used during the Great Patriotic War.[11] The contemporary utility of transmitting disinformation through exercises has been confirmed in the restricted General Staff journal *Voennaya mysl' (Military Thought)*. In discussing approaches to establishing control over the decision-making and actions of potential opponents, Major General M. Ionov argued that this may be achieved by, among other things, "deluding the enemy as to one's intentions, capability, state, and actions of troops."[12] Ionov singled out the use of strategic-scale exercises as especially useful for this purpose. Obviously, one objective of such an approach could be to persuade NATO decision-makers that any resort to nuclear weapons would be met by an all-out unlimited response. Equally obvious, the fact that the Soviets have more recently been urging the West to watch their exercises for evidence of the defensive nature of the new Warsaw Pact military doctrine suggests they have been convinced that the West was monitoring and drawing intelligence conclusions from Warsaw Pact exercises.

Because no command is told whether its actions are part of a deception plan and each is required to prepare its own deception plan in support of its combat actions, commands are never aware of whether they are preparing a deception plan to a higher level command's deception plan or are protecting the real operations. Such practice does not inhibit adequate preparation of troops, since the objective is development of a capability to conduct combat actions in any environment and under any conditions. This process includes restricting the exercised use of weapons and equipment to mislead the enemy further as to Soviet capabilities. Even this knowledge aids the Soviet effort because the West must still question the accuracy of the information it has been able to obtain.

The success of these Soviet efforts is evident in many Western analyses of Soviet military strategy. One American assessment, for example, concluded:

> Any U.S. or NATO limited nuclear strike that is not large enough to eliminate USSR superiority in both conventional and nuclear arms in Eurasia might well be ignored by the Soviets.... Conversely, NATO nuclear strikes large enough to deny the USSR its objectives would probably be viewed by the Soviets as the prelude to massive nuclear attacks by all available Western forces. In this case, the USSR would preempt the attack with its own massive power in strategic and operational-tactical systems (or respond with its own massive power if it did not succeed in preempting completely). In both scenarios, U.S. or NATO limited nuclear operations would be ineffectual or (more probably) would bring about the very catastrophe they were designed to avoid.[13]

This conclusion has become an important component of the Western critique of NATO strategy. The view that reliance on nuclear threats in the current strategy is seriously flawed is based, in part, on the assertion that NATO and the Warsaw Pact hold fundamentally different assumptions regarding the controllability of nuclear warfare. Consequently, this conclusion has contributed to the increasing uncertainty over the credibility of NATO's threat to use nuclear weapons and the overall viability of the flexible response strategy. More important, such a conclusion influences the peacetime context of Western military planning by undercutting the acquisition of capabilities to provide a wider range of options from which NATO decision-makers could select to restrain Soviet pursuit of wartime strategic objectives.

Efforts to discredit NATO strategy and undercut the rationale for future procurements necessary to maintain flexible response intensified in the aftermath of the 27th Congress of the Communist Party of the Soviet Union (CPSU) in February 1986. The Soviet Union has sought to persuade the West that, with the advent of "new political thinking," it no longer regards military power as the main tool for fulfilling national security objectives. Henceforth, according to Soviet spokespersons, the Soviet Union will rely increasingly on political and diplomatic means of attaining those objectives. As a result of the revolutionary nature of many recent changes in the Soviet Union, even the most ardent Western skeptics have adopted a wait-and-see attitude toward the evolution of Soviet national security policy. NATO faces a number of critical decisions regarding its force posture at the turn of the century. As a result, to the extent that a wait-and-see policy is accompanied by moratoriums on such decisions,[14] the new Soviet approach could prove to be an effective tool for slowing the momentum of Western force modernization programs.

The problem, of course, is that the debate over the role of military power in meeting Soviet national security objectives has not been concluded. Senior Soviet military spokespersons do not seem persuaded that the military instrument has somehow lost its utility in the future management of East-West competition. Deputy Chief of the Soviet General Staff General M. A. Gareev has pointedly reminded the political leadership that "the harsh lessons of the past attest to the fact that an abstract policy does not and cannot exist in a pure form."[15] Gareev dismisses Soviet civilian strategic theorists for their failure to comprehend that "in the real international situation the struggle for peace and the readiness to defend it with weapons in hand do not contradict but complement each other."[16]

Consequently, while it is clear that the West must respond to the emerging changes in the Soviet Union, it must evaluate the strategic environment on the basis of Soviet operational concepts and their ability to execute these concepts. Despite force reductions and changes in Soviet exercise patterns, such adjustments are not inconsistent with the traditional Soviet predilection for offensive operations. Soviet proponents of a nonoffensive operations posture have, to date, had only marginal impact on the operational planning of the Soviet General Staff.

Soviet Wartime Strategic Objectives

Considerable confusion has also developed in recent years over Western interpretations of Soviet strategic objectives. Some have asserted that these objectives have undergone a number of shifts since the mid-1960s but have generally continued to focus on "not losing" a future war in Europe.[17] More recently, such analysts have argued that Soviet objectives are oriented solely to "securing the territorial integrity and internal cohesion of the Soviet bloc."[18] Although they would certainly agree that such goals represent overriding political objectives, Soviet military planners would find it puzzling to see such goals interpreted as strategic objectives. By definition, strategic objectives are "the planned final result of military actions on a strategic scale, the achievement of which will lead to a large, sometimes sharp change in the military-political and strategic situation, and which would make possible the successful conduct of the war and its victorious conclusion."[19]

To "avoid losing" would hardly seem to fit into the Soviet definition of strategic objectives. Army General I. E. Shavrov, in the lecture on military strategy at the Voroshilov General Staff Academy cited earlier, listed the "total destruction of the enemy's armed forces and his military economy, neutralization of his state control system, and seizure of his territory" as specific strategic objectives to be pursued in the event of a future war.

More recent sources reflect continuity in Soviet assessments of the nature of wartime strategic objectives. These objectives continue to be identified as the defeat and destruction of enemy forces, the disruption of the enemy's economic potential, the disorganization of the enemy's systems of state and military control, and occupation of enemy territory.[20]

Direct attacks on NATO forces and strategic command, control, communications, and intelligence capabilities would be coupled with coercive efforts designed to disintegrate NATO cohesion and force individual members out of the war. Such strategic objectives would facilitate the political objective of minimizing the economic costs to the Soviet state. To this end, the Soviets would avoid destruction of Western Europe's rich industrial base in order to exploit this potential fully in its own recovery and to avoid burdening itself further with supporting the populations of occupied territories. In this sense, Soviet strategic objectives regarding enemy economies are, in the case of a strategic operation in a continental TSMA, focused on disruption rather than destruction.[21] Thus, Soviet objectives against enemy economies would focus on the disruption and disorganization of their ability to supply and support military operations.

The defensive-only thrust of the military policy espoused by some Soviet strategic theorists continues to be soundly rejected by military planners in favor of traditional Soviet wartime strategic objectives. In major doctrinal statements, for example, Defense Minister Army General D. T. Yazov has emphasized the requirement to destroy enemy armed forces.[22] Similarly, Colonel General M. A. Gareev has continued to call for the "full destruction of the enemy" in the event of a future war.[23] Both agree that this strategic objective may be fulfilled only through the execution of "decisive offensive operations," which, in contrast to Soviet advocates of new political thought, indicates that disorganization and disruption of the enemy's strategic control and seizure of enemy territory continue to guide Soviet operational planning by the General Staff.

At the same time it should be recognized that, while these might be Soviet strategic objectives at the outset of a European war, such objectives could be subject to redefinition or modification. Political objectives and considerations are the overriding factors in determining the specific content of Soviet strategic objectives. The primacy of political objectives is one constant within the Soviet theory of military strategy.

Soviet flexibility with regard to strategic objectives is facilitated by the fact that the Soviet theory of military strategy provides for limited strategic objectives: "Limited strategic objectives could be to annihilate an armed forces group of the enemy in a theater or a certain sector of a theater, to destroy the economic and military potential of one or several enemy nations, to disrupt the state administrative and war directing systems of the enemy, to remove from the war one or several nations of an enemy alliance, etc."[24] Conceptualization of limited strategic objectives in theater warfare provides a basis for redefining military and political objectives, especially if the Soviets were in danger of losing control over the course of a conflict. The resurrection of "partial victory" as an important principle in Soviet strategic theory, after it had been discarded in the 1960s, may be a further indication of Soviet recognition of the need for flexibility in the event of future wars.[25]

Similarly, there has been little change in the basic operational concepts for attaining Soviet strategic objectives. Since the early 1930s, the conduct of deep operations has been held by Soviet military theoreticians to be the most effective method of achieving rapid, decisive successes in theater warfare.[26] The centrality of this concept, which envisions the simultaneous extension of both fire destruction and maneuvering across the entire depth of the enemy's deployments, has remained a constant within Soviet planning for theater warfare. Until the advent of the nuclear missile, however, deep operations objectives could be achieved only sequentially by the combined-arms activities of Soviet *fronts* and armies. With the widespread deployment of nuclear missiles, Soviet military planners had a weapon that could be used to influence the enemy directly and immediately throughout the depth of their entire territory, eventually including even overseas territories. Massive strategic nuclear strikes into the deep interior meant Soviet strategic objectives in the TSMA could be achieved in a matter of hours or even minutes, whereas in the past, weeks and even months were required.

For reasons discussed in more detail later in this chapter, this represented little more than a passing, poorly devised phase of Soviet military development. The risks associated with such options and the opportunities inherent in flexible response led the Soviets back to a more traditional approach to deep operations. By the mid-1970s, although they expected the tempo of these operations to be much higher, they also realized that because of the limited capabilities of available conventional weapons systems, fire destruction of the enemy could be achieved only "successively (not simultaneously)."[27] The introduction of longer range, highly accurate conventional fire systems, however, seems to hold out the possibility that simultaneous fire destruction is once again within reach.

Consequently, although the basic strategic concept of deep operations has changed little, the methods and means of executing this concept have changed dramatically. Moreover, Soviet military theoreticians have recently been forecasting another change in methods to reflect the growing convergence between offensive and defensive means of armed conflict. The main factors behind these changes have been the introduction of new technologies, the evolving strategy and forces of potential opponents, the "objective" assessment of the Soviet military's own capabilities, and ongoing changes in the objective conditions that provide the context for military planning. One such change in the objective conditions—and the starting point for many of the Soviet force development efforts of the past twenty years—was the sober assessment of the realities of nuclear war.

The Role of Nuclear Weapons in Theater Warfare

In his attempt to substitute firepower for manpower and the nuclear missile weapon for general purpose forces, Khrushchev denied himself and the Soviet state flexibility by reducing the number of available choices in the use of military power. According to S. A. Tyushkevich, "After the October [1964] Plenum of the CC CPSU, certain incorrect views within military-scientific circles connected with the over-evaluation of the potential of the atomic weapon, its influence on the character of war, and on the further development of the Armed Forces were overcome."[28]

While Soviet military planning continued to focus attention on an enemy's first mass nuclear strike as the most stressing scenario, Soviet military theoreticians concluded that the emerging correlation of strategic nuclear forces was eroding U.S. willingness to execute such a strike.[29] There was simply "too great a risk of the destruction of one's own government and the responsibility to humanity, for the fatal consequence of nuclear war is too heavy for an aggressor to make an easy decision on the immediate employment of nuclear weapons from the very beginning of a war without having used all other means for the attainment of its objectives."[30] The Soviets sought to reinforce this emerging strategic situation by acquiring forces that provided both damage limitation—to be achieved with preemptive nuclear strike capabilities and active and passive defensive measures—and a secure reserve to ensure retaliation in the event of a successful U.S. first strike. In the view of Soviet military planners, only the ability to thwart the objectives of a first use of nuclear weapons by the United States would have any hope of restraining such use.

It is doubtful the Soviet military ever had much confidence in their ability to prevent widespread destruction to the Soviet homeland in the event of general nuclear war and, over time, the Soviets became even more pessimistic about the consequences of such a war. Lecture materials from the Voroshilov General Staff Academy explain that "in such circumstances, victory would be achieved by the side that manages to inflict heavy damage on the enemy, while retaining its economy and armed forces to a greater extent and rapidly restoring the combat capabilities of its armed forces."[31] The stark consequences attending this victory, however, were reflected in the grim conclusion that "the characteristics of a nuclear war lead us to conclude that in a nuclear war there will be no winner or loser."[32]

Beyond the disincentives to be found in the consequences of a general nuclear war, the Soviets became increasingly skeptical about the utility of large-scale nuclear strikes in theater operations. Once NATO nuclear forces became sufficiently diversified to survive Soviet nuclear preemption, nuclear war threatened to introduce more problems than it resolved. Residual NATO nuclear forces would be sufficient, in the Soviet view, to disrupt a Soviet advance effectively. The consequences of this operational dilemma were explained in the following manner:

> As a result of the influence of nuclear weapons, whole subunits, units and even formations can lose their combat capabilities within a few minutes. Large territories will become useless for immediate continuation of the operation. Combat actions along many axes will become isolated and local in nature, troop control will be interrupted, and because of the break in the originally outlined plans for combat actions, new decisions will have to be made within restricted time limits.[33]

Even the most cursory familiarity with Soviet military planning literature is sufficient to realize that such a combat environment is an anathema to the Soviet commander and his staff. As a result of a simultaneous exchange, the most favorable nuclear scenario Soviet planners could now assume was that

> much time will be required for the restoration of units' combat effectiveness, eliminating the impact of the enemy's nuclear blows, waiting for the reduction of the degree of radiation and reconstruction of roads and column routes.
> The process of restoring the combat capabilities of units after the mutual initial nuclear strike and the organization of units to launch the offensive in compliance with the new situation might take one or two days or even more. Consequently, it is advised that for practical purposes the rate of advance in offensive operations should be planned and considered identically for immediate and subsequent objectives in both nuclear and nonnuclear operations.[34]

Thus, in a combat environment wherein the enemy maintained a nuclear retaliatory capability, nuclear weapons offered no solution to the problem of an increasing density of antitank weapons because nuclear weapons could not necessarily counteract the decelerating effect of the proliferation of relatively inexpensive antiarmor technology. As a result, very early in the post-Khrushchev period, some Soviet military theoreticians were arguing that "for one of the sides, which has achieved the necessary results and is successfully developing the offensive, it will be advantageous to delay the beginning of the use of nuclear weapons as long as possible."[35]

Although the Soviets believed that the threat of nuclear war had been reduced, they understood that it had not been eliminated entirely, and they continued to be appropriately skeptical about their ability to prevent a theater war from escalating into a nuclear war. In a retrospective analysis of Soviet military theory of the 1960s and 1970s, General Gareev indicated that most Soviet military works of the period "proceeded from the conclusion that war in all circumstances would be conducted with the use of nuclear weapons, and strategic military actions with the use of only conventional means of destruction were considered as a short episode at the beginning of the war."[36] While the Soviets initially believed that NATO would be forced to resort to nuclear use early (three to five days at best), this estimate was modified somewhat during the 1970s. The Soviets observed that NATO's nuclear threshold during this period, as evidenced in NATO exercises, was raised to five to seven days and later possibly even to ten days and beyond.[37]

Consequently, although the Soviets clearly preferred conventional operations themselves, their confidence that a war could remain conventional throughout was not really reflected in authoritative statements of military strategy or military doctrine until the late 1970s or early 1980s. For example, Marshal N. V. Ogarkov's 1979 entry on "Military Strategy" in the *Soviet Military Encyclopedia* is indicative of how far the Soviets had moved away from the 1960s mass nuclear employment scenarios and pessimism over the ability to control escalation. According to Ogarkov, Soviet military strategy assumes that world war may begin and continue for a certain time with the use of only conventional weapons. However, the expansion of strategic military actions *may* lead to its transformation into a general nuclear war, the main means of the conduct of which will be strategic nuclear weapons." [Emphasis added.][38]

In 1982 the Soviets included a similar formulation in military doctrine. Military doctrine, by definition, represents the Party's guidance to the military on such subjects as the nature of future war, the probable enemy or enemies, and the military-technical tasks of the armed forces, including the methods of conducting such a war. According to a statement of the conclusions of military doctrine published that year, "a future war may be unleashed with either conventional or nuclear weapons; having begun with the use of conventional weapons it may at a certain stage be transformed into a nuclear [war]."[39] Acceptance of this formulation into Soviet military doctrine was apparently based on the Party leadership's conclusion that "the successes of the Soviet Union in the area of military technology and weapons have convinced the imperialist strategists both of the doubtfulness of their concept of the shattering destruction of the USSR by means of a surprise massive nuclear strike and of the inevitability of retaliation."[40]

From this it is evident that by the early 1980s, the Soviet military and political leadership had both agreed that the only inevitability about the introduction of nuclear weapons into a conflict was retaliatory strikes in response to an attempted surprise first strike. The verbal formulations of both military strategy and military doctrine seemed to reflect a Soviet conclusion that the threshold of nuclear use in a future conflict was likely to be quite high and perhaps might never be reached at all.

This conclusion was, however, subject to some reconsideration during the mid-1980s. The early Reagan administration's rhetoric and published reports that the administration's first Defense Guidance Plan focused heavily on requirements for conducting a protracted nuclear war led to a resurgence of references to the inevitability of escalation to nuclear weapons and concern over the form and content of the initial period of war.[41] It has also been reported in the West that Soviet concern was sufficiently high to necessitate the declaration of a KGB alert in 1981.[42]

As compelling as these concerns may have been, there can be little doubt that the war scare of the early 1980s was also part of the internal political maneuvering in preparation for the impending Soviet leadership succession. That one aspect of this maneuvering apparently involved the military's attempt to exploit the war scare in order to squeeze resource allocations out of the Party is indicated by the Party's denunciation of those who overestimated the threat of global nuclear war.[43] The Soviet affair is also difficult to separate from the public diplomacy campaign designed to disrupt deployment of U.S. INF missiles in Europe and to dissuade NATO from modernizing its nonnuclear forces. Marshal Kulikov's argument that "by whatever means a new world war begins it will inevitably end in a nuclear catastrophe"[44] should, therefore, be seen at least partially as an attempt to exploit the general level of concern over nuclear war prevalent at that time in the West. A key feature of this campaign was the Soviet assertion that the threshold between conventional and nuclear operations was becoming increasingly blurred by the impending introduction of newer conventional systems that would inevitably increase the threat of war and thereby of nuclear catastrophe.

The war scare was short lived, and by 1985 renewed Soviet confidence in the U.S. acceptance of the futility of large-scale nuclear use was reflected in two authoritative statements of the contemporary views of Soviet military theory. In a February 1985 article in the authoritative Party journal *Kommunist*, the then Chief of the General Staff, Marshal S. F. Akhromeev, concluded:

> In recent years our probable enemies, recognizing the unavoidability of a retaliatory nuclear strike and its catastrophic consequences, are devoting special attention to the development of systems of conventional weapons with better destruction, range, and accuracy characteristics. Simultaneously,

they are modernizing the methods of unleashing strategic military actions with the use of conventional means of destruction, primarily the new types of controlled and automated modes of high-accuracy weapons.[45]

That same year General Gareev, then Chief of the General Staff's Military Science Directorate, argued that earlier assumptions about escalation no longer prevailed in Soviet military theory. Gareev concluded that the Soviet proponents of these assumptions had simply failed to foresee that the "accumulation and modernization" of the nuclear arsenals on both sides would reach such levels as to ensure that "a mass use of these weapons in a war could bring catastrophic consequences for both sides."[46] Consequently, nuclear deterrence of nuclear use, coupled with Western (and Soviet) efforts to develop highly accurate conventional systems, led Soviet military theoreticians to conclude that the "possibility of a comparatively long war with use of conventional weapons is increasing."[47] In other words, despite the war scare and Soviet references to the increased threat of nuclear war in the early 1980s, by 1985 the chief military scientist of the Soviet Union could argue that the Soviets believed the nuclear threshold of a future conflict was likely to be so high it might never be reached.

Soviet estimates of the potential of future conventional systems suggest they believe that the advent of these new technologies will raise the nuclear threshold of a major war. Marshal Ogarkov essentially argued that the integration of new reconnaissance and guidance technologies into future conventional weapons systems should effectively remove any limitations on the application of these systems to a wider range of missions.[48] Such improvements could result in conventional weapons with an effectiveness approaching that of nuclear weapons, make these systems "global in nature," and sharply increase their destructive potential "at a minimum by an order of magnitude."[49] Obviously, such a conclusion supports the military's claim for their traditional share of resources, but it also reflects a genuine desire to avoid both the risks and the planning imponderables of operations in a nuclear environment.

Retention of a set of effective, selective nuclear options, however, does ease the potential dilemma associated with response to a limited NATO nuclear use. The same logic that led to the conclusion that nuclear use does not resolve conventional operation problems when an opponent possesses a secure nuclear capability also argues against unrestrained use of such weapons. To expect somehow that once nuclear weapons have been used on a selective, limited basis, the disincentive associated with large-scale, indiscriminate use would no longer determine the nature of the Soviet response, would violate the Soviet view of the "logic of war." The Soviet political and military leadership would continue to insist on attempting to control the course of such a conflict; nothing would threaten such control more effectively than a large-scale, indiscriminate exchange of nuclear weapons.

Resort to such limited nuclear options would not be made lightly, however. It seems likely, given the clear perception by both the political and military leadership of the potential risks associated with any nuclear use, that such options would be considered in only the most desperate circumstances. Equally important, the existence of secure and effective theater nuclear forces, and especially the threat to use such a force in an irrational manner, could deter an opponent from using its nuclear forces. Consequently, the continued existence of such nuclear forces has a real military utility in the sense that its deterrent capability might enable the Soviets to execute a nonnuclear strategic operation unimpeded by an opponent's nuclear attack.

Although the observation that the Soviets believe it would be possible to restrict nuclear operations to geographic regions (such as TSMAs) has generally been accepted among American intelligence analysts, the possibility that the Soviets would impose restrictions on strikes within these regions flies in the face of conventional wisdom. The Soviets clearly concluded some time ago, however, that both their planning process and their force posture must be responsive to a broad range of targeting requirements in different scenarios. Early in a conflict, for example, the target set might be quite large, although both political factors and considerations of military effectiveness could impose limitations against even this target. As explained in the Voroshilov General Staff Academy lecture materials, given Soviet intentions to postpone the initiation of nuclear operations for as long as possible, "as the attacking elements of the fronts advance more and more in depth, most of the enemy targets initially selected to be engaged by friendly nuclear weapons will be seized or destroyed."[50] Obviously, as the number of surviving enemy targets decreased, the Soviet nuclear strike would be sized accordingly.

Beyond this rational process of sizing strikes, the Soviets themselves concluded some time ago that both their planning process and their force posture must be sufficiently flexible to meet the demands of unanticipated restrictions on nuclear use. General Shavrov's lecture on military strategy, for example, explicitly acknowledged that

> in some TSMAs or in vast areas of one TSMA different conditions may prevail with regard to the restriction on the use of nuclear weapons. We can assume that in some cases the nuclear strikes will be limited to the military and economic targets deployed in smaller areas, while in other conditions nuclear strikes may be limited only to military targets such as first echelon forces. Moreover, other forms may develop which are beyond our estimation at this point.[51]

Soviet ground forces spokespersons, in particular, continue to underscore the training requirements associated with the conduct of limited nuclear operations.[52]

Preferred Soviet Approaches to Achieving Strategic Objectives

When the Soviets expected a future conflict in Europe to become nuclear quickly, the requirements were clear and simple. The Soviets saw the competing strategies to be essentially a struggle for the initiative. The operational concept was to conduct the breakthrough with nuclear weapons and to exploit it with tank-heavy mechanized forces. Once it was understood that a future battlefield might never become nuclear, old problems reappeared and new solutions had to be found.

The possibility of a nonnuclear conflict in Europe brought back defense as an operational form, required the ability to mass fires without massing the means in conducting breakthroughs, and increased both the potential scale of nonnuclear operations and the potential intensity of such a conflict. With regard to the latter, the Soviets believed that future nonnuclear conflicts would be characterized by an intense struggle to seize the initiative, to impose one's will on the opponent, and to dictate the contours of the conflict. A key objective would be to keep an opponent continuously off balance through unanticipated modes of fire and methods of maneuver.[53]

In effect, the Soviets believed that the combination of massed but controlled firepower and maneuver could substitute for nuclear strikes. The Soviets have never accepted the simple-minded assertion that such a substitution could somehow be accomplished on a one-for-one basis, nor do they assume such substitutions to be feasible even with the introduction of precision-guided munitions. Instead, the Soviet approach has been to substitute the concentrated and focused application of firepower and maneuver of several services and arms of service, in a combined-arms fashion, at the decisive places and times in order to achieve similar operational impact.

In response to the increasing scale of future operations, the Soviets turned to the concept of a strategic operation in a continental TSMA, which would coordinate the activities of a number of Soviet *fronts*, as a substitute for the single *front* exploitation of nuclear strikes envisioned in the 1960s.[54] The key components would be the air operation, operations by *fronts*, an air defense operation, assault landing operations, and naval operations.[55] Each operation would have its own objectives, timelines, and specific methods but would be conducted according to a unified plan under the overall control of a single commander.

To provide for flexible unified control, in 1984 the Soviets established High Commands of Forces in the western and southwestern TSMAs. The Soviets had already taken steps, however, to fully integrate Warsaw Pact forces directly into the Soviet command structure. Between 1979 and 1980, all Warsaw Pact states except Romania signed a "status of the Combined Armed Forces and Wartime Command Organs" agreement.[56] In the case of Poland, at least, this meant that "the Soviet commands will give their orders and instructions directly to the Polish armies subordinate to them, bypassing the national command. In practice this means that the USSR has the unrestricted right to dispose of the Polish People's Army without any prior consultation with the [Polish] leadership."[57]

These efforts to improve Soviet control were combined with a modernization and restructuring program between 1978 and 1982 that has been described as resulting in the fielding of a force capable of "conducting combat operations both with the use of nuclear weapons and with the use of only conventional means."[58] The Soviets' confidence in their accomplishments over this period was indicated by Lieutenant General M. Kir'yan's 1982 conclusion that "a well-proportioned military organization has been created, which would permit the accomplishment of missions of any scale, under any conditions."[59] Thus, the Soviet military seemed cautiously confident with regard to at least their near-term potential to attain Soviet strategic objectives in the event of a major European war.

One unique feature of contemporary warfare, however, is the growing convergence of offensive and defensive means. Warsaw Pact military theorists concluded that emerging technologies could enable the defender to seize the initiative, change the correlation of forces in its favor, and create optimum conditions for counteroffensive operations.[60] As a result, Soviet military theorists have concluded that to speak of offensive and defensive operations "in their pure forms" is no longer meaningful. Even in defensive operations, for example, the Soviets anticipate the conduct of offensive combat operations. Such an integrated approach would seem well suited to overcoming the potential effects of attrition, particularly on forces operating on the main direction. This principle would appear to apply equally to forces operating on operational or strategic axes, and possibly even to entire TSMAs.[61]

Given the primacy of the western TSMA and the potential imposition of political constraints on mobilization during the threatening period, the Soviets might opt for strategic defensive operations on the flanking northwestern and southwestern TSMAs. These operations would be designed to hold NATO forces in place and inhibit their concentration in the central region to defeat the strategic offensive in the main TSMA. This Soviet military geographic approach to strategic planning appears to leave the Soviets with what, in all practicality, amounts to a horizontal as well as a vertical view of escalation. Horizontal escalation, like vertical escalation, however, is not seen to be a policy but merely an event. Although the Soviets fully appreciate the need to prepare various options for simultaneous executions in multiple TSMAs, the Soviet policy objective is to provide for victory in the main TSMA while restraining the enemy from escalation as a rational strategic alternative to defeat.

With the initiation of the strategic offensive in the main TSMA, maneuvering forces would attempt to engage and hold the stronger NATO corps while trying to execute deep penetrating envelopments through the weaker NATO corps. The tactical effect would be to engage all of NATO's first-echelon corps in battle, which would mask the Soviet strategic plan. The operational intent would be to confuse "echelons above corps" by having each corps commander report a defense against the main attack. To the extent NATO committed its reserves in the wrong places, NATO commanders would find themselves gaining tactical "victories" while being encircled at the operational and operational-strategic scale.

Consequently, the Soviets believe the battlefield of the near-term future would be one of mutual penetration by both offense and defense. In such an environment, survival and success would depend on "the capability of conducting attacks at a high tempo, for rapidly taking advantage of the results of casualties inflicted by firepower, and by simultaneously improving their protection against massive casualties by the enemy's weapons"[62] through maneuver. The Soviets are also convinced that combat activities "will have a land and air nature, which means that victory will be directly dependent on the conditions of the air situation, the stability of the air defense system, and high effectiveness in the use of the air forces."[63]

The air situation would be critical because aviation is estimated to constitute approximately 50 percent of NATO's firepower in the central region in both a nuclear and nonnuclear conflict.[64] The Soviets would seek to shift radically the correlation of means through suppression of NATO's air power at main operating bases and to gain a sufficient correlation of forces to penetrate the enemy's defenses successfully by concentrating strike groupings. By defending elsewhere at a less than 1:1 ratio of forces, the Soviets would hope to obtain a 5:1 ratio in the main sectors. The operational objective is to exploit the penetration in order to encircle NATO's first-echelon corps so that they could not be withdrawn to reestablish a defensive line, which would require yet another penetration phase of the offensive.

The Soviets consider penetration to be the single most critical phase of the offensive. It is then that the offensive force is most vulnerable to attack, and, therefore, the penetration phase is also the most costly in forces and means. During World War II, the breakthrough phase of army operations, despite being characterized by a 4:1 to 7:1 ratio of superiority, accounted for 65 to 80 percent of any given army's losses over the course of its offensive.[65]

In December 1981, Colonel-General F. Gaivoronskii, the deputy commandant of the Voroshilov General Staff Academy, summarized the major elements for success in this phase:

> Under present-day conditions when only conventional weapons are used in the conduct of the offensive, effecting a penetration could require the application of large masses of artillery, aircraft, and tanks; thorough suppression of the enemy's numerous anti-tank weapons; and protection of the attacking forces from air strikes, including strikes by combat helicopters. To increase the tempo of the offensive and stop the approach of enemy reserves towards the penetration sectors, it will be necessary to make air and missile strikes throughout the depth of the enemy's defenses and make wide use of airborne (air assault) landings.[66]

In an effort to facilitate the penetration phase, the Soviets modified the initiation of fire support so that it begins when the penetrating forces depart their forward assembly area and continues even after the supported forces have completed their penetration. Of the four phases of offensive fire support only the first constitutes something new in the 1980s:

- Fire support of the movement from the departure area to the line of deployment (beginning perhaps more than an hour before the supported force reaches the forward edge of the battle area [FEBA]).
- Fire preparation of the assault (beginning 20 to 30 minutes before the FEBA is reached).
- Fire support of the assault (beginning when the assault or movement into the penetration actually begins).
- Fire accompaniment of the attacking troops (this phase will continue in the enemy rear area until the mission is completed).

Once NATO's tactical zone of defense (the first 30 to 50 kilometers) has been penetrated, follow-on forces would be directed at the successful penetration. This effort would include an attempt both to widen the shoulders of the penetration and quickly to move as much force as possible into the depths of the defense. As other divisions and armies follow the operational mobile groups (OMGs) into the depths of NATO defenses, these forces would attempt rapidly to widen the arms of encirclement and, subsequently, to develop the offensive on an external front.[67]

Since the Soviets believe the cost of penetration does not decrease once a mean level of superiority (usually 7:1) is reached, and since force-to-space ratios would not allow a *front* to deploy all its resources securely, second-echelon armies of first-echelon *fronts* could be hundreds of kilometers from the FEBA at the commencement of hostilities. Such armies would move forward to be 200 to 300 kilometers from planned commitments. These forces would proceed as divisions, and possibly as armies, by road with their full air defenses in place. In avoiding movement by rail once full hostilities have begun, follow-on forces would be more capable of defending themselves against deep attack from NATO air power. Road movement would also allow for more flexible responses to the evolving situations. Thus, second-echelon *fronts* would not be seen to move forward but would appear as new commands established on the external front and grow in size over time as new forces were made available.

Beyond the military implications of operational success, the Soviets would hope to obtain greater political leverage against smaller states in the NATO coalition. The Soviets are likely to try to enhance the political leverage of operational success with a direct assault on the only small front-line state, Denmark. A successful effort to coerce Denmark out of the Western coalition would quickly be followed by an intensification of similar efforts against the Netherlands and Belgium.[68]

Soviet Views on Long-Term Competition

Despite the successes of the past 20 years, the Soviet military leadership remains fully aware of the requirement to look continuously to the future. In fact, the Soviet military science community seems well along in its assessment of the impact and implications of emerging military technologies for future warfare. The combination of a number of trends has led the Soviets to forecast potentially significant changes in the nature of theater warfare over the next two decades. Not only are the Soviets factoring in those attributes of emerging military technologies that they consider likely to produce qualitative transformations in the nature of future warfare; they are also examining the implications of the efforts by both NATO and the Warsaw Pact to develop new operational concepts and force structures for using the new technologies more effectively. Although they anticipate the emergence of these developments over the next decade, the Soviets also recognize that these are susceptible to funding and resource allocation decisions.

First, by all indications, the Soviets expect the scale of military operations to increase progressively over the coming decades. Soviet military theoreticians believe that the attributes of future nonnuclear weapons will enable these systems to perform expanded missions not only at tactical, but also at operational and perhaps even strategic depths. Similarly, the Soviets believe they need to prepare for increasingly prolonged theater conflicts. Soviet military theoreticians appear to see no contradiction between this conclusion and their view that these systems would increase the tempo of conventional operations dramatically. They have apparently concluded that the combination of anticipated attrition, interdiction of forward deploying forces, disruption of control, and the increasing complexity of operations under these conditions may produce prolonged conflicts. Consequently, regardless of the means used, Soviet military theoreticians expect it will be "necessary to be prepared for a long, stubborn, and bitter armed struggle."[69]

Closely related to this theme are evolving Soviet views on the future relationship between offense and defense. For reasons of sound military theory and operational necessity, by the early 1980s the Soviet military had already accepted the requirement to strike a more harmonious balance between offensive and defensive operations in future military planning. The impending introduction of high-accuracy, long-range systems and the development on both sides of operational strategies for their employment have led Warsaw Pact analysts to undertake a "complete reevaluation of the very essence of the defense on the future battlefield."[70]

Warsaw Pact analysts have concluded, as part of this reevaluation, that "the difference between offense and defense is obliterated."[71] No longer will the defender be forced to cede the initiative to the attacker, nor will the defender's range of action be limited simply to the tactical zone. Instead, the defender will be able to strike deeper into the enemy's rear at the most advantageous time. According to one Soviet military theoretician:

> These weapons with great range are definitely forcing a greater dispersal of troops which are located in depth. Regions of concentration, assembly regions for the offensive, etc., will have to be larger. Reserves, second echelons, and forces approaching from the rear will have to be shifted and rearranged in smaller formations over greater areas so as to make of themselves the greatest possible number of potential targets.[72]

Beyond these considerations, Warsaw Pact military theoreticians anticipate changes in the basic trends in force development in the following areas:

- Traditional land operations will be transformed into air-land operations; this trend is already well under way but is expected to accelerate with more widespread use of special combat helicopters.
- The role and importance of mobility will increase for all military operations.
- The practice of combat operations within enemy formations, especially raiding and OMG actions, is expected to develop further and be widely disseminated among the forces.
- Battle will be initiated at increasingly greater distances.

- The "information struggle" to tighten control over one's own forces and impose control over enemy forces through disinformation and other techniques is growing in significance.[73]

These changes will drive not only decisions concerning armament, but also the methods of operations and the structure of the forces themselves. As Major-General Vorob'yev predicted in a Soviet *Military Thought* article in 1980, once armies are equipped on a large scale with these qualitatively new means of armed struggle, the current phase in military development will give rise to the emergence of a new race between offensive and defensive means of war and the formulation of new concepts of military strategy.[74]

Notes

1 Research for this chapter was completed in December 1988. An earlier version of this paper is scheduled for publication in Bruce Parrott, ed., *The Dynamics of Soviet Defense Policy*.
2 Other translations frequently used in the West are "Theater of Military Operations" (TMO) and "Theater of Military Action" (TMA). For our reasons for preferring TSMA, see John G. Hines and Phillip A. Petersen, "Changing the Soviet System of Controls," *International Defense Review*, 3 (1986), pp. 281–289.
3 V. M. Kulish, ed., *Voennaya sila i mezhdunarodnye otnosheniya [Military Force and International Relations]* (Moscow: Voenizdat, 1972), p. 222.
4 For Western skepticism on the "hostage Europe" explanation, see Matthew A. Evangelista, "Stalin's Postwar Army Reappraised," *International Security* (Winter 1982/83), pp. 110–138. Evangelista's interpretation is unconvincing for several reasons. Evangelista asserts that "the Red Army simply did not have the capability to execute the rapid invasion feared by many in the West. As late as 1950, half of its transportation consisted of horse-drawn vehicles. Most roads, railways, and bridges in Eastern Europe and the Soviet Union were destroyed or seriously damaged during the war." Evangelista apparently overlooked the fact that during the Vistula-Oder campaign of the Great Patriotic War, for example, Soviet forces advanced 500 kilometers in 21 days with horse-drawn vehicles over a transportation network destroyed by retreating German forces in a vain effort to slow the Soviets.
5 See Arnold L. Horelick and Myron Rush, *Strategic Power and Soviet Foreign Policy* (Chicago: University of Chicago Press, 1966), p. 94.
6 Conversation of Marshal P. A. Rotmistrov, June 4, 1965.
7 Yu. Nepodaev, "On the 'Nuclear Threshold' in NATO Strategy," *VM*, 6 (1966); translated in *FBIS FPD* 0503/67, May 26, 1967, p. 72.
8 "Military Strategy: Lectures from the Voroshilov General Staff Academy," *The Journal of Soviet Military Studies*, 1 (1988), p. 46.
9 Although we now have strong evidence to support the conclusion that the Soviets have, since 1945, effectively used Western strategic theorists to reinforce the Soviet capacity to dissuade the West from specific policies, this tendency toward willful ignorance in support of particular policy preferences has not gone unnoticed by some Western theorists. See, for example, George H. Quester, "On the Identification of Real and Pretended Communist Military Doctrine," *The Journal of Conflict Resolution*, 6 (1966), pp. 172–179.
10 N. V. Ogarkov, *Istoriya uchit bditel'nost' [History Teaches Vigilance]* (Moscow: Voenizdat, 1985), p. 89.

11 Examples of Soviet deceptive practices during the war are well documented in Western sources. See, for example, David M. Glantz, "The Red Mask: The Nature and Legacy of Soviet Military Deception in the Second World War," *Intelligence and National Security*, 7 (1987), pp. 175–259; Notra Trulock III, "The Theory, Planning, and Practices of Operational Deception," in Brian Daley and Patrick Parker, eds., *Soviet Strategic Deception* (Boston: Lexington Books, 1987). *The Journal of Soviet Military Studies*, 6 (1988), pp. 262–275, provides details of an early Soviet deception and a Soviet General Staff critique of its effectiveness. Soviet sources on the topic are numerous, including Marshal of the Soviet Union (MSU) G. K. Zhukov, *Vospominaniya i razmyshleniya [Reminiscences and Reflections]*, Vol. I (Moscow: Novisti, 1975), pp. 172–174.

12 Maj. Gen. M. Ionov, "On the Methods of Influencing an Opponent's Decision," *VM*, 12 (1971), as translated in *Selected Readings from Military Thought 1963–1973*, Vol. 5, Part II (Washington, DC: Government Printing Office, 1975), p. 167.

13 The quotation is from William T. Lee and Richard F. Staar, *Soviet Military Policy since World War II* (Stanford, CA: Hoover Institute Press, 1986), pp. 39–40. It is consistent, however, with the interpretations expressed by other Western military Sovietologists. See, among others, Benjamin S. Lambeth, "Selective Nuclear Operations and Soviet Strategy," in Johan J. Holst and Uwe Nerlich, eds., *Beyond Nuclear Deterrence* (New York: Crane, Russak & Co., 1977), pp. 91–93; Michael MccGwire, *Military Objectives in Soviet Foreign Policy* (Washington, DC: The Brookings Institution, 1987), p. 81; and Stephen Meyer, *Soviet Theater Nuclear Forces*, Part I, Adelphi Paper No. 186 (London: International Institute for Strategic Studies, 1983), p. 30. More even-handed treatments of this issue can be found in Fritz W. Ermarth, "Contrasts in American and Soviet Strategic Thought," in Derek Leebaert, *Soviet Military Thinking* (London: George Allen & Unwin, 1981), p. 62; and Robbin F. Laird and Dale R. Herspring, *The Soviet Union and Strategic Arms* (Boulder, CO: Westview Press, 1984), pp. 75–83.

14 See, for example, Townsend Hoopes, "Let Us Forge Plowshares in Central Europe," *International Herald-Tribune*, September 3–4, 1988.

15 Col. Gen. M. A. Gareev, "Great October and Defense of the Motherland," *Oktyabr'*, 2 (1988).

16 Ibid.

17 See the discussion of Soviet strategic objectives in MccGwire, *Military Objectives*, pp. 36–59.

18 Michael MccGwire, "New Directions in Soviet Arms-Control Policy", *The Washington Quarterly* (Summer 1988), p. 187.

19 Marshal S. F. Akhromeev, *Voennyi entsiklopedicheskii slovar' [Military Encyclopedic Dictionary]* (Moscow: Voenizdat, 1986), p. 313.

20 See, for example, Maj. Gen. N. N. Kuznetsov, "Strategic Objectives," *SVE*, Vol. 7 (Moscow: Voenizdat, 1981), p. 552.

21 See, for example, Col. M. Shirokov, "Military Geography at the Present Stage," *VM*, 11 (1966); translated in *FBIS FPD* 0730/67, July 23, 1967.

22 Army Gen. D. T. Yazov, *Na strazhe sotsializma i mira [On Guard of Socialism and Peace]* (Moscow: Voenizdat, 1987), p. 33.

23 Col. Gen. M. A. Gareev, *Sovetskaya voennaya nauka [Soviet Military Science]* (Moscow: Znaniye, 1987), p. 36.

24 Maj. Gen. N. N. Kuznetsov, "On the Categories and Principles of Soviet Military Science," *VM*, 1 (1984), as translated in the Vietnamese journal *Tap Chi Quan Doi Nhan Dan*, 6 (1984).

25 Col. Gen. M. A. Gareev, *M. V. Frunze: voennyi teoretik [M. V. Frunze: Military Theoretician]* (Moscow: Voenizdat, 1984), p. 241.

26 See, for example, I. A. Korotkov, *Istoriya sovetskoi voennoi msyl'—kratkii ocherk. 1917–yun' 1941 [The History of Soviet Military Thought—Short Outline. 1917–June 1941]* (Moscow: Nauka, 1980), pp. 147–148.

27 "Army Offensive Operations," *The Voroshilov Lectures: Materials from the Soviet General Staff Academy, mid-1970s* (Washington, DC: National Defense University Press, 1989).

28 S. A. Tyushkevich, *Sovetskie Vooruzhennye Sily: istoriya stroitel'stva [The Soviet Armed Forces: History of Development]* (Moscow: Voenizdat, 1978), p. 476.

29 See, for example, Maj. Gen. N. Vasendin and Col. N. Kuznetsov, "Modern Warfare and Surprise Attack," *VM*, 6 (1968); and Army Gen. S. Ivanov, "Soviet Military Doctrine and Strategy," *VM*, 5 (1969); both translated in *Selected Readings*.

30 Ivanov, *VM* 5 (1969), p. 28.

31 "Principles of Strategic Action of the Armed Forces," *The Voroshilov Lectures*.

32 "Military Strategy," *The Journal of Soviet Military Studies*, 1 (1988).

33 I. G. Zav'yalov, "The New Weapon and Military Art," *KZ*, October 30, 1970; translated in William F. Scott, ed., *Selected Soviet Military Writings, 1970–1975* (Washington, DC: Government Printing Office, 1977), p. 209.

34 "Army Offensive Operations," *The Voroshilov Lectures*.

35 Col. D. Samorukov, "Combat Operations Involving Conventional Means of Destruction," *VM*, 10 (1967); translated in *Selected Readings*, Part I, pp. 174–175.

36 Gareev, *Frunze*, p. 240.

37 See, for example, Col. Gen. M. A. Gareev, *Obshchevoiskovye ucheniya [Combined Arms Exercises]* (Moscow: Voenizdat, 1983), p. 225.

38 MSU N. V. Ogarkov, "Military Strategy," *SVE*, Vol. 7, p. 541.

39 M. M. Kir'yan, *Voenno-tekhnicheskii progress i Vooruzhennye Sily SSSR [Military-Technical Progress and the Armed Forces of the USSR]* (Moscow: Voenizdat, 1982), p. 312.

40 Ibid., p. 313.

41 On the inevitability issue, see, for example, MSU V. Kulikov, "To Curb the Arms Race," *KZ*, February 21, 1984, p. 3. On increased attention to operations in a nuclear environment, see Col. F. D. Sverdlov, *Takticheskii Manevr [Tactical Maneuver]* (Moscow: Voenizdat, 1982); and Army Gen. P. N. Lashchenko, *Iskusstvo voenachal'nika [Art of the Military Commander]* (Moscow: Voenizdat, 1986). On the initial period of the war, compare Maj. Gen. V. Matsulenko, "Certain Conclusions from the Experience of the Initial Period of the Great Patriotic War," *VIZ*, 3 (1984), with Lt. Gen. A. I. Yevseev, "On Certain Trends in the Changes of the Content and Character of the Initial Period of War," *VIZ*, 11 (1985).

42 Murrey Marder, "Defector Told of Soviet Alert, KGB Station Reportedly Warned U.S. Would Attack," *Washington Post*, August 8, 1986, p. 1; and Gordon Brook-Shepherd, *The Storm Birds: Soviet Post-War Defectors* (London: Weidenfeld and Nicholson, 1988), p. 267. The authors wish to express their appreciation to John Wood for obtaining a copy of this volume for them when doing so was difficult.

43 Indications that the war scare was exploited to manipulate resource allocation is provided in Jeremy R. Azrael, *The Soviet Civilian Leadership and the Military High Command, 1976–1986* (Santa Monica, CA: RAND Corporation, R-3521-AF, June 1987), pp. 22–37. While the KGB defector Oleg Gordievskii, the original source of the information that "the Western World had passed through a war danger zone," admits that the nature of the crisis was questioned even in the Soviet Union, he believes it was genuine. See Brook-Shepherd, *The Storm Birds*, pp. 267–271.

44 Kulikov, *KZ*, February 21, 1984, p. 3.

45 MSU S. F. Akhromeev, "The Superiority of Soviet Military Science and Soviet Military Art—One of the Most Important Factors of Victory," *Kommunist*, 3 (1985), p. 62.

46 Gareev, *Frunze*, p. 240.

47 Ibid.
48 These conclusions were stated most explicitly in MSU N. V. Ogarkov, "The Defense of Socialism: The Experience of History and the Contemporary Period," *KZ*, May 9, 1984, p. 3.
49 Ibid.
50 "Military Strategy," *The Journal of Soviet Military Studies*, 1 (1988), p. 46.
51 Ibid.
52 See, for example, the articles by the current Soviet Ground Forces Deputy CinC for Combat Training, Col. Gen. Merimskii, and the former Chief of the Ground Forces Rocket Troops and Artillery, Marshal Peredelskii: Col. Gen. V. A. Merimskii, *Takticheskaya podgotovka motostrelkovykh tankovykh podrazdelenii [Tactical Preparation of Motorized Rifle and Tank Subunits]* (Moscow: Voenizdat, 1984), p. 8; and Marshal Arty G. E. Peredelskii, *Otechestvennaya artilleriya 600 let [600 Years of the Homeland's Artillery]* (Moscow: Voenizdat, 1986), p. 329.
53 See, for example, Col. Gen. D. Brinkevich, "The Factor of Time in Battle," *VV*, 11 (1986); and the detailed analysis of these methods in M. M. Kir'yan, *Vnezapnost' v nastupatel'nykh operatsiyakh Velikoi Otechestvennoi Voyny [Surprise in the Offensive Operations of the Great Patriotic War]* (Moscow: Nauka, 1986).
54 For single-front exploitation, see S. N. Kozlov, M. V. Smirnov, J. S. Baz', and P. A. Siborov, *O sovetskoi voennoi nauke [On Soviet Military Science]* (Moscow: Voenizdat, 1964).
55 Lt. Gen. P. A. Zhilin, *Istoriya voennogo iskusstva [History of Military Art]* (Moscow: Voenizdat, 1986), p. 400.
56 Ryszard Jerzy Kuklinski, "The War against the Nation Seen from the Inside," *Kultura*, Paris No. 4/475 (1987); translated by Defense Intelligence Agency, No. LN 678-87, June 18, 1987, p. 43.
57 Ibid., p. 44.
58 Kir'yan, *Voenno-tekhnicheskii progress*, p. 326.
59 Ibid. For the best overview of the 1978–1982 Soviet force reorganization, see Soviet Army Studies Office, *The Soviet Conduct of War* (Ft. Leavenworth, KS: SASO, 1987), pp. 8–13. For the best overview of the current Soviet reorganization, see Christopher Donnelly, *Red Banner: The Soviet Military System in Peace and War* (London: Jane's Information Group, 1988).
60 Col. Stanislaw Koziej, "Anticipated Directions for Change in Tactics of Ground Troops," *Przeglad Wojsk Ladowych*, September 1986, p. 4. The authors are indebted to Dr. Harold Orenstein and Graham Turbiville, both of the U.S. Army's Soviet Army Studies Office, the former for his translation and the latter for bringing this article to their attention. See also Maj. Gen. I. N. Vorob'yev, "The Relationship and Reciprocal Effects Between Offense and Defense," *VM*, 4 (1980); translated in the Vietnamese journal *Tap Chi Quan Doi Nhan Dan*, 1 (1982); translated by Defense Intelligence Agency, No. LN-893-86, September 9, 1986.
61 Compare the entry under "Strategic Defense" in MSU N. V. Ogarkov, ed., *Voennyi entsiklopedicheskii slovar' [Military Encyclopedic Dictionary]* (Moscow: Voenizdat, 1983), p. 710, with the same entry in the second edition, MSU S. F. Akhromeev, ed., *Voennyi entsiklopedischeskii slovar'*, p. 710.
62 Vorob'yev, *VM*, 4 (1980).
63 Ibid.
64 Col. Gen. M. Zaitsev, "Organization of Air Defense—An Important Mission of the Combined Arms Commander," *VV*, 2 (1979), p. 23.
65 Maj. H. F. Stoeckli (Swiss Army), "Soviet Operational Planning: Superiority Ratios vs. Casualty Rates," study published by the Soviet Studies Research Centre, Royal Military Academy, Sandhurst, p. 5.

66 Col. Gen. F. Gaivoronskii, "The Development of Operational Art," *VIZ*, 12 (1981), pp. 28–29.
67 For more detail on encirclement operations, see John G. Hines, "Soviet Front Operations in Europe—Planning for Encirclement," *Spotlight on the Soviet Union*, a report from a conference at Sundvollen, Norway, April 25–27, 1985. Report 1/86 (Oslo: Alumni Association of the Norwegian Defense College, 1986), p. 88.
68 For the best discussion of Soviet views on opposing coalitions, see John J. Yurechko, *Coalition Warfare: The Soviet Approach* (Koln: Bundesinstitut fur ostwissenschaftliche und internationale Studien, 1986). For Soviet intentions against Denmark, see Christopher N. Donnelly and Phillip A. Petersen, "Soviet Strategists Target Denmark," *International Defense Review*, 8 (1986), pp. 1047–1051.
69 Gareev, *Frunze*, p. 241.
70 Koziej, *Przeglad Wojsk Ladowych* (September 1986), p. 4.
71 Col. Stanislaw Koziej, "Is There Only Hope?" *Zolnierz Wolnosci*, May 13, 1988. The authors are indebted to Dr. Harold Orenstein of the Soviet Army Studies Office for bringing this article to their attention and for his translation. See also Vorob'yev, *VM*, 4 (1980).
72 Zaitsev, *VV*, 2 (1979), p. 3.
73 Koziej, *Przeglad Wojsk Ladowych* (September 1986), p. 9.
74 Vorob'yev, *VM*, 4 (1980), p. 9.

3
Gorbachev and the Western Alliance: Reassessing the Anticoalition Strategy[1]

Erik P. Hoffmann
State University of New York, Albany
Institute for East-West Security Studies

Mikhail Gorbachev's administration has thoughtfully reassessed the USSR's traditional anticoalition strategy toward the West. Unexpected international developments and persistent domestic economic problems and political conflicts have complicated this reassessment. Whether Soviet foreign policy professionals have conceptualized and are carrying out a new anticoalition strategy is a moot point. But one can see the beginnings of a modified strategy or, at the very least, a more sophisticated and differentiated implementation of the traditional strategy.

The anticoalition strategy of Nikita Khrushchev, Leonid Brezhnev, and Gorbachev has consisted of efforts to reduce U.S. influence in Western Europe and to forestall the creation of a strong and independent Western Europe. The chief goals of this strategy are to maximize the USSR's security by deterring an attack and military threats from the West; to hinder Western, especially U.S., attempts to fragment or destabilize the Soviet bloc; to encourage "realistic" thinking, especially about nuclear and conventional arms control, among U.S. and West European leaders and citizens; and to accelerate Soviet and East European economic modernization by increasing the trade of technology with the West.

Joseph Stalin's successors have become increasingly sensitive to the shifting policies and institutional relationships of the Western Alliance. Soviet analysts have come to see these changes as cyclical rather than unidirectional.

Khrushchev, Brezhnev, and Gorbachev have all used classic divide and conquer tactics and have made serious attempts to improve relations with their superpower adversary, the United States, and their key European rival, the Federal Republic of Germany (FRG). These different approaches have been pursued sometimes sequentially and sometimes simultaneously. However, the traditional anticoalition strategy of Khrushchev and Brezhnev placed considerable emphasis on expanding fissures between Western and Eastern Europe. Hallmarks of this strategy were efforts to develop long-term diplomatic and commercial ties between the USSR and individual West European countries; to weaken European Economic Community (EEC) integration and limit cooperation between the EEC and the Council for Mutual Economic Assistance (CMEA); and to undermine NATO integration and abolish both the Warsaw Pact and NATO.[2]

Gorbachev and his closest party-government supporters have striven to shape the security and economic policies of key countries in the Western Alliance more than to fragment its integrated structures, especially NATO and the EEC. Arms control agreements with broad support in the West (such as the Strategic Arms Limitation Talks, SALT I and II) and commercial deals involving many nations (for example, the Soviet–West European natural gas deal) were considered to be highly desirable if these policies were responsive to Soviet concerns. Unified Western policies on strategic and economic rivalry (for example, the Strategic Defense Initiative [SDI] and Coordinating Committee for Multi-lateral Export Controls [COCOM] restrictions) were considered to be highly undesirable if these policies were unresponsive to Soviet concerns. And Western disunity on military, commercial, and human rights issues was considered to be detrimental if it prolonged mutually disadvantageous East-West security, economic, and political relations.

Top Soviet leaders have traditionally viewed U.S.-Soviet summits as important means of influencing Western policies. Summits are especially favored by general secretaries striving to consolidate their power, and Gorbachev needed summit accomplishments to help reduce domestic resistance to his restructuring of society. Also, the Gorbachev administration immediately took a vigorous and multidirectional approach to East-West relations, exploring ways to influence the U.S. president both directly and indirectly through his European allies. Gorbachev's generation of leaders is more realistic than Brezhnev's about Soviet leverage vis-à-vis Western Europe and about West European leverage vis-à-vis the United States. And Gorbachev's Politburo has concluded that it can fruitfully negotiate with the dominant leader of the Western Alliance, however intransigent that person may temporarily be, while trying to weaken but not dismantle this surprisingly cohesive coalition.

Gorbachev and other top Soviet officials have emphatically denied that they are placing "wedges" between the United States and its West European allies. But the USSR has often exacerbated fissures within and among the nations of the Western Alliance. Under Gorbachev countless efforts have been made to portray the United States as a bully vis-à-vis its Western allies; to affirm that the United States and West European nations have many different interests and that the USSR and West European nations have many similar interests; to exhort the "realistic" leaders and citizens of Western Europe to resist the "hegemonic" aspirations of the United States and the "vassallike" submissiveness of their compatriots; to contend that the United States seeks military superiority over the USSR and sacrifices European security to achieve this end; and to debunk the "myths" that the USSR seeks military superiority over the United States and threatens Western Europe with the Warsaw Pact's strategic or conventional forces.

While reiterating such traditional perspectives, the Gorbachev administration has added some new emphases. Soviet spokespersons have voiced heightened concern and frustration with the West European leaders' reluctance to resist U.S. "intimidation" on security issues; with the West European politicians' and industrialists' eagerness to reap commercial gains from participation in the Reagan administration's SDI program; and with the periodic slippage of Soviet diplomatic, economic, and military leverage over West European nations despite energetic and innovative Soviet efforts to increase this leverage.

Soviet confidence in its West European policies rose enormously with the treaties perpetuating the USSR-FRG rapprochement of the late 1960s and early 1970s and was reinforced by the divergent U.S. and West European responses to the Afghan and Polish crises beginning in 1979 and 1980 respectively. But Soviet policy was undermined by the installation of U.S. intermediate-range nuclear forces (INF) in Western Europe in 1984 and by the ebbing of the West European peace movements. Soviet puzzlement, anxiety, and anger intensified in 1985 and 1986 with the conservative West European governments' acquiescence in a series of U.S. military actions (for example, in Grenada and Libya), indifference to Gorbachev's "peace offensive" (such as moratoriums on nuclear explosions), and support for nuclear weapons after both the United States and the USSR provisionally agreed to substantial reductions at the Reykjavik summit (for example, West European rejection of the "zero option").

Briefly stated, Gorbachev and his colleagues have contended that the major West European countries now possess enhanced capabilities to defend their own and pan-European interests within the Western Alliance but their leaders lack the political will to use these capabilities in a constructive, sustained, and coordinated manner. Through private and public diplomacy,

Soviet officials have coped with such setbacks as key West European governmental and corporate support for SDI and INF deployment. However, the Gorbachev administration has exhibited a growing understanding of the fissures in the Western Alliance as well as of the military-security, scientific-technological, and socioeconomic bonds that link the United States and Western Europe.

Gorbachev and his colleagues have experimented with diverse methods to reduce the West's cohesiveness and to cooperate with a unified Western Alliance. The exacerbation of "interimperialist contradictions" was the cornerstone of the traditional anticoalition strategy but has been downgraded to the role of a selectively used tactic. Placing wedges in the Western Alliance was not an end in itself and, until summitry has fulfilled or exhausted its potential, may no longer be the chief means of achieving larger ends. Rather, wedges and summitry were among the increasing number of instruments that Gorbachev simultaneously and serially utilized in the pursuit of military detente, which he views as the sine qua non of further political and economic detente.

How does one identify the Gorbachev administration's evolving perspectives on Western Europe and ascertain whether they comprise, reflect, or portend a new anticoalition strategy? The term *perspective* is used in the broadest sense to include images, perceptions, expectations, motivations, strategies, goals, priorities, and tactics. Images are articulated verbally and in writing, but perceptions, expectations, motivations, strategies, goals, priorities, and tactics must be inferred from words and deeds. Most important, one attributes meaning to Soviet public pronouncements and international behavior by assessing their congruence with a given theory—in the present case, the theory that the Gorbachev Politburo is revising the USSR's traditional peacetime and wartime anticoalition strategy vis-à-vis the Western Alliance.[3]

One must, however, distinguish between Soviet declaratory and actual policies and try to explain the purposes and effects of these policies and their interrelationships. From the content of Soviet public declarations one must infer the propaganda strategy or goal; from the propaganda strategy or goal one must infer the elite strategy, priorities, and tactics; from the elite strategy, priorities, and tactics one must infer elite expectations and perceptions; and from elite expectations and perceptions one must infer international and domestic situational factors and capabilities.[4]

Even in highly stable political systems it is difficult to identify and order the key variables cited and to confirm generalizations about propaganda strategy and goals (declaratory policy) and elite strategy, priorities, and tactics (actual policy). It was especially challenging to identify and order these variables in the first years of the Gorbachev administration, because leading foreign policy personnel were rapidly being replaced and vital institutions and institutional relationships were being altered or adjusted. Also, Gorbachev and other top officials affirmed that they were espousing "new political thinking," "breaking away from stereotypes and prejudices," and "gradually learning a lot" and that "international relations are too complex [and] have too many layers and too many facets for us to be able to settle in advance on any one method of doing business."[5]

Gorbachev, in a private meeting with a group of Soviet writers, is reported to have said: "Our enemy has figured us out. They are not frightened of our nuclear might. They are not going to start a war. They are worried about one thing: if democracy develops under us, if that happens, then we will win. Therefore, they have started a campaign against our leadership, using all available means, right up to terror. They write about the apparatus that did in Khrushchev and that will now do in the new leadership."[6] Vladimir Lomeiko, special Soviet envoy and head of the delegation to the Vienna Conference on Security and Cooperation in Europe (CSCE), declared in an interview with Austrian journalists: "We have repeatedly said that we do not want to drive a wedge between the allies. For one simple reason: For us it is much easier to deal with a sensible, united will of the West instead of with different, even opposite opinions."[7]

Such pronouncements reveal more about Soviet propaganda strategy and actual policy, about policymaking procedures and policy alternatives under consideration, than many Western analysts acknowledge. Why? Because Gorbachev has made known to domestic and international audiences more of actual Soviet policy and the calculations and logic behind it. Gorbachev and his international affairs advisors have been more forthcoming than their predecessors about domestic/international linkages, about the strategic and tactical underpinnings of their policies, and even about their emotional reactions to the consequences of recent initiatives. Also, some Soviet leaders and analysts have made a renewed effort to understand Western perspectives on international politics. They consider misperceptions of one's adversaries to be a mutual problem and imply that accurate perceptions are necessary but not sufficient to produce greater trust or cooperation. Although the Gorbachev administration's interpretations contain logical inconsistencies and factual distortions and lacunae, reasoned argumentation and presentation of evidence are becoming the norm rather than the exception.

Greater congruence between declaratory and actual foreign policies, together with increasing debate between conservatives and reformers, have characterized Soviet media commentary on international politics under Gorbachev. Soviet conservatives (for example, Andrei Gromyko) have emphasized East-West conflict and West-West cooperation, whereas Soviet reformers (for example, Gorbachev) have emphasized East-West cooperation and

West-West conflict. To be sure, many conservatives were removed or cowed during the overhaul of the foreign policymaking institutions and staffs in 1986. But Gorbachev has frequently bemoaned the deep-seated conservative resistance to his reorganization of Soviet policymaking and administration and has stressed the close interconnections between domestic and foreign policies.

There were stark differences between an ultraconservative "platform" supported by Politburo member Yegor Ligachev in March 1988 and an ultrareformist response by Politburo member Aleksandr Yakovlev a month later. Although open debates were nearly as sharp at the subsequent 19th conference of the Communist Party of the Soviet Union (CPSU), Gorbachev chose to view these debates as important manifestations of public candor (*glasnost'*) and restructuring (*perestroika*) in the party-state. He did not view with magnanimity Ligachev's and Yakovlev's major public disagreements on foreign policy in the summer of 1988, and Ligachev was stripped of his responsibilities for supervising ideological activities in the domestic and international spheres. However, because Gorbachev values the clash of most interests and opinions and because the principle of *glasnost'* is being applied increasingly to foreign policy as well as domestic policy issues, diverse perspectives on the past, present, and future will surely continue to be criticized, praised, tempered, and evaded in highly politicized contexts at home and abroad.

Experimentation has been a central feature of the Gorbachev administration's policy toward Western Europe, and difficulty in understanding developments and choosing among options have been central dilemmas. Top Soviet officials have hinted that the traditional anticoalition strategy has been of diminishing effectiveness, but that some present-day adjustments, innovations, and prescriptions are proving insufficiently effective too. Quite possibly, influential Soviet reformers have agreed to pursue a much more flexible, differentiated, and vigorous anticoalition strategy toward the West. Such an approach would consist largely of exploring new ways to reward West European governments that independently support Soviet interests or that pressure the United States to do so. A revitalized Soviet search for ways to punish recalcitrant West European governments, including stepped-up espionage and more active support for opposition parties and movements, could be anticipated if Soviet leaders concluded that their constructive initiatives had been repeatedly rebuffed or ignored. But Gorbachev and his colleagues have proffered carrots rather than brandished sticks. They continue to use wedges, but much less consistently and indiscriminately than did their predecessors, probably in the hope of reemphasizing the cooperative rather than the confrontational components of peaceful coexistence and of reviving the East-West detente of the early 1970s. Although belittling Brezhnev is now obligatory in Soviet propaganda strategy, Gorbachev is well aware that he has yet to match Brezhnev's achievements with Richard Nixon and Willy Brandt and that the 27th Party Congress policies will founder unless he does so.

Under Gorbachev, more accurate and insightful Soviet perspectives on Western Europe have emerged. Soviet declaratory policy toward the West has been a dynamic mix of praise and criticism, and actual policy has been a dynamic mix of cooperation and confrontation. Inducements and threats have been used to keep Western Europe as balkanized as possible and to make key countries less willing and able to unite with one another and with the United States, thereby jeopardizing Soviet interests. But the Western countries most willing to further Soviet interests are not the countries most able to do so or most important to overall Soviet strategy. The United States and the FRG are the most able and important but initially were the least willing under conservatives Ronald Reagan and Helmut Kohl. Herein lay a basic challenge to Gorbachev—one that he confronted directly and met effectively.

This chapter scrutinizes some of the major components of Soviet declaratory and actual policies toward the Western Alliance, with emphasis on the shifting combinations of continuity and change. Particular attention is devoted to analysis of the Gorbachev administration's perspectives on Western Europe; reassessment of the objectives and effectiveness of the traditional anticoalition strategy; and political, economic, and military responses to SDI and INF. In conclusion, the likelihood of significant changes in Gorbachev's anticoalition strategy is evaluated by an examination of domestic and international pressures and constraints.

Perspectives on Western Europe

Gorbachev and his Politburo colleagues are pursuing active rather than passive coexistence with the West. They affirm and apparently believe that East-West relations are not a zero-sum game. In other words, they think that mutually beneficial political, military, and economic relations between the USSR and the United States and between the USSR and Western Europe can be achieved simultaneously and are not mutually exclusive. According to the highly authoritative 1986 CPSU program, a chief goal of Soviet foreign policy is "to maintain and develop the USSR's relations with capitalist states on the basis of peaceful coexistence and businesslike, mutually advantageous cooperation."[8] Or, as *Izvestiya* foreign affairs commentator Aleksandr Bovin concluded, "Anyone trying to build up his own security will ultimately aggravate his own military-strategic security."9

Significantly, Gorbachev and his Politburo colleagues perceive Soviet-U.S. relations to be by far the most important component of Soviet foreign policy. Because Western Europe broadly supports detente and the Reagan administration for a long time rejected it, Gorbachev values stronger ties between the USSR and the large and small nations of Western Europe and between the nations of Western and Eastern Europe. Soviet analysts view the international system as increasingly multipolar but still dominated by relations between the two superpowers. To be sure, they characterize the United States, Western Europe, and Japan as three competing capitalist centers and the rest of the world as increasingly interdependent. But they contend that the U.S.-Soviet military and ideological rivalry preserves the essence of the post-World War II bipolar system.

These general Soviet perspectives on Europe are reflected in the 1986 CPSU program[10] and in numerous pronouncements by Gorbachev. For example, he has stated that "Europe's historic chance, its future, lies in peaceful cooperation of the states of that continent. And it is important, while preserving the capital that has been built up, to move forward from the initial phase of detente to a more stable, mature detente, and then to the creation of reliable security on the basis of the Helsinki process and of radical cuts in nuclear and conventional arms."[11]

The responses of Deputy Foreign Minister Anatolii Adamishin to a Vienna *Kurier* journalist illustrate one high-ranking Soviet policymaker's view of the cohesion within NATO as well as the enhanced importance of Western Europe:

> *Kurier*: Since Gromyko left, Soviet foreign policy has seemed to concentrate less on America alone....
> *Adamishin*: From the technical point of view you are right: We do not want to see the world with American eyes any longer. This is a new approach. But it has nothing to do with Gromyko.
> *Kurier*: Has Europe become more important for you?
> *Adamishin*: Yes, we have discovered Western Europe as an independent power—and we are facing it more actively, dynamically, and openly. More contacts, more talks, more ideas. But very often we encounter a Western Europe that says: "Better talk with America." Obviously you are still too much ashamed of your own interests.[12]

As Adamishin's final remarks indicate, the Gorbachev administration was distressed by European deference to U.S. interests and assessed Soviet relations with Western Europe partially in instrumental terms.

Soviet leaders have long affirmed that the USSR can and should influence the United States's West European allies, which in turn can and should influence the United States. This strategy or important tactic has been clearly articulated by Soviet analysts. For example, Bovin calls for reduced U.S. influence in Western and Eastern Europe and claims that the USSR is not striving to eliminate the U.S. presence in Western Europe. He implicitly acknowledges that the goal of eliminating U.S. influence in Western Europe is neither feasible nor desirable. It is unfeasible because of the abundant and deep-seated political, military, economic, and cultural ties among the major capitalist nations. It is undesirable because of Western Europe's much stronger bargaining position and uncertain policies vis-à-vis the Soviet bloc. Moreover, Bovin finds lesser goals both feasible and desirable. West European nations have the capability to influence U.S. policy on key security, commercial, and scientific-technological issues and must be induced to do so. Bovin declared:

> Soviet policy takes into account the differences of views between Western Europe and the United States. But it does so by no means in order to squeeze the United States out of Europe and gain political control of the continent which it so longs for, in the opinion of "perspicacious" analysts in the West. Our objective is much more modest. We would like to utilize Western Europe's potential to make good, via the transatlantic channel, the obvious shortage of common sense in the incumbent U.S. Administration.... Since Western Europe and the United States are allies, and in our opinion the elements of common sense in European politics are stronger, we are attempting to get Western Europe to influence the United States in order to make American policy more sober, reflecting modern quality to a greater extent. To a certain extent the Europeans are doing this, but in my view they could do more.[13]

Conversely, Soviet observers contend that U.S. and West European efforts to influence the USSR by influencing its East European allies must be resisted by better coordination of the Soviet bloc's foreign policy. Although Gorbachev has propounded such views, they are the hallmark of conservative Soviet officials and commentators. For example, O. Vladimirov asserts:

> Imperialism, not abandoning its "crusade" against communism, is gambling on the export of counterrevolution and on direct interference in the socialist states' affairs and is attempting to play on their "specific character" and "special role." The aim is the same—to weaken the alliance of fraternal countries, to alienate and isolate them from the USSR, and ultimately to attempt to secure an erosion and even a change of social system. This line is pursued in a differentiated fashion, rather subtly, by a combination of threats and promises and, as its architects cynically admit, on the basis of the socialist countries' "conduct." Roles are appropriately allocated among the United States and its NATO allies. International opportunism, religious centers, and forces laying claim to hegemonism implement their political and propaganda line with regard

to the community countries in an equally differentiated fashion (pursuing their own objectives).[14]

Some East European leaders and most citizens want to stabilize and integrate the diplomatic, security, and economic ties between Eastern and Western Europe while making them much less dependent on the vicissitudes of the U.S.-Soviet rivalry. But the Gorbachev administration, like all of its predecessors, has resisted such efforts to bifurcate the USSR and Eastern Europe and to undermine Soviet control in Eastern Europe. Likewise, some West European leaders and citizens, especially social democrats, want to stabilize and integrate the diplomatic, security, and economic relations between Eastern and Western Europe while decoupling them from the superpower rivalry. But the Gorbachev administration has resisted such efforts too.[15]

Soviet leaders in the post-Stalin period have increasingly stressed that all European nations—including the USSR—have many similar interests (such as nuclear-free zones and expanded trade) and many interests that differ from those of nations on other continents, especially the United States (for example, Europe's "bridgehead" role in an East-West nuclear exchange and trade embargoes). Politburo members have often maintained that persistent U.S. attempts to impede cooperation between Western Europe and the Soviet bloc must be resisted by West European and Soviet bloc nations. But the Gorbachev administration has placed less emphasis on the ideological differences between the "socialist" and "capitalist" nations of Europe and more emphasis on their geographic, cultural, and political bonds. Soviet spokespersons claim that West and East European countries share a "common home" and "one all-European civilization" and can and must function as an autonomous political body, cooperating to enhance pan-European interests and Europe's importance in international politics.

The following authoritative Soviet statements illustrate the tactic of driving wedges into the Western Alliance by stressing the common heritage and concerns of the USSR and Eastern and Western Europe as well as European dissimilarities with the United States. Gorbachev proclaimed:

> The ancient Greeks have a myth about the abduction of Europa. This fairy tale subject has unexpectedly acquired contemporary content. It goes without saying that, as a geographical concept, Europe will remain where it is. But one gets the impression that the independent policy of some Western European states has been kidnapped and is being transported across the ocean, that under the pretext of protecting security, both the national interests and destinies of the 700-million-strong population of our continent, and the civilizations that have been developed here since time immemorial, are being sold off.[16]

Fyodor Burlatskii, the prominent political commentator, affirmed: "All of Europe today is inseparably linked by a common nuclear fate. The Europeans are perfectly aware that not a single one of them will manage to survive in the event of a nuclear war. This probably applies to the entire world, but it applies mainly to us Europeans. And if this is so ... then the Europeans ... must take their fate into their own hands."[17] Vadim Zagladin, Central Committee member and first deputy head of its International Department, stated:

> The successful development of the all-European and the Helsinki processes is above all disturbed by forces that are not of a European nature. One does not have to be a foreign political expert to notice that the United States does everything it can to slow down or hinder entirely relations between the Eastern and the Western part of Europe. For this reason they plan to take both political and economic measures. The COCOM restrictions aim not so much at hindering the transfer of military technologies to the East (anyway, our countries have proved that they are able to solve any kind of problems as regards national defense), but rather at slowing down and curbing commercial, economic, technological, and scientific exchange between the two parts of the continent.[18]

In a word, the Gorbachev administration has pursued with new vigor the goals of convincing West European politicians and citizens that the superpower arms race and the increasingly interdependent regional economy are giving them more interests in common with the socialist countries of the Soviet bloc and fewer interests in common with their capitalist superpower ally across the Atlantic.

Reassessing the Objectives of the Traditional Anticoalition Strategy

In determining the USSR's objectives toward Western Europe, the Soviet leadership has always considered its principal enemy to be the nation that was temporarily dominant in Europe. Until 1924 Soviet policy was mainly anti-French, until 1930 mainly anti-British, until 1939 mainly anti-German, and after 1945 mainly anti-American.

Without conducting a rigorous content analysis of the Soviet media, one cannot be certain about the extent to which present-day Soviet officials characterize the United States as an enemy. Although they rarely use the term *enemy* to describe the United States, they sometimes do so when describing imperialism, or imperialist nations.[19] To be sure, reformist leaders especially have urged capitalist and socialist nations to discard enemy images of one another. But conservative leaders especially have viewed the large and small West European countries as victims of or misguided accomplices in "imperialist" activities instigated by the United States. And the Gorbachev administration certainly identifies the United States as the chief obstacle to greater Soviet influence in Western Europe.

Angela Stent identifies five long-standing Soviet aims vis-à-vis Western Europe:

> There has been remarkable continuity in Soviet policy toward Western Europe since World War II, involving five basic objectives: ... to contain West Germany and control East Germany ...; to encourage fissures within the Atlantic alliance ...; to prevent a more coherent political, economic, and military Western European integration ...; to assist the growth of communism within Western Europe ...; and to import large amounts of Western European technology and equipment.[20]

John van Oudenaren affirms that

> While promoting Western Europe's transition to socialism remains an ultimate Soviet objective, a number of near-term objectives that bear directly on the state interests of the USSR have greater significance for policy. These objectives include: 1. Safeguarding the Soviet Union's World War II territorial and political gains from internal or external challenge. 2. Gradually lessening the American military, political, economic, and cultural presence in Western Europe. 3. Obtaining a voice in the defense policies of Western European countries. 4. Securing economic and technological inputs for the Soviet economy. 5. Obtaining leverage over the internal politics and policies of Western European countries, particularly on matters that affect Soviet interests. 6. Hindering progress toward Western European unity under European Community (EC) or other auspices.[21]

The Gorbachev administration seems to be pursuing these traditional objectives enumerated by Stent and van Oudenaren. The present-day Soviet anticoalition strategy is based in large part on the following assumptions, all of which were operative under Brezhnev and to a lesser extent under Khrushchev: First, a weak Western Alliance is not possible because of the strong U.S.–West European cultural, economic, and political-military ties, nor is it desirable because of the likelihood of creating a much more independent and united Western Europe that might pursue assertive economic and military policies toward the USSR and Eastern Europe. Second, a strong Western Alliance is possible for short time periods chiefly because of U.S. "blackmail" and "crude pressuring" and West European "relinquishing of sovereignty" and "pandering," but it is desirable only if the West's unity is voluntary and if key Western and Soviet economic and military policies are congruent. Third, a divided Western Alliance is highly likely because of enduring "interimperialist contradictions," and it is highly desirable as a means of influencing West European economic and military policies through bilateral ties, thereby reducing the concentration of U.S.-West European or intra-West European power that could endanger the USSR in times of peace, crisis, or war. Fourth, stronger bilateral and multilateral ties between countries of the Soviet bloc and Western Europe are quite possible because of many common European interests, and such ties are quite desirable as a means of improving prosperity and security throughout Europe and of strengthening pan-European influence throughout the world, especially vis-à-vis the United States's "hegemonistic" East-West and North-South economic and military policies.[22]

Like Khrushchev's and Brezhnev's anticoalition strategies, Gorbachev's strategy has minimal and maximal goals. Minimally, Soviet leaders have tried to preserve fissures created by the Western allies themselves. Maximally, Soviet leaders have tried to create and expand new fissures. The Soviet leadership considers moderate disagreement within the Western Alliance to be a much more stable condition than either a high degree of unity or disunity. Bilateral ties between the USSR and major West European nations are considered especially important, and bilateral ties between and multilateral ties among major West European nations are considered especially threatening, unless these ties are for the purpose of furthering both Soviet and West European interests (for example, regional nuclear-free zones and the Soviet-West European natural gas deal).[23]

The Gorbachev leadership has striven especially hard to alter the policies and, to a much lesser extent, the institutions of the Western Alliance. The weakening of institutions is not viewed primarily as an end in itself but as a means of altering Western policies. Western political-military and commercial policies are, in turn, viewed as a means of furthering or obstructing the USSR's most significant goals—national security, political stability, economic modernization, and control over Eastern Europe. Gorbachev insists that "we want to have good relations not only with Western Europe but also with the United States" and that "we are realists and we understand how strong are the ties—historical, political, and economic—linking Western Europe and the United States."[24] Georgii Arbatov, a Central Committee member and leading Soviet specialist on the United States, concluded a televised exchange with British politician David Owen as follows: "So we are not for splitting your alliance but for seeing that alliance concluding a proper policy. If that does happen, please be friendly with the United States."[25]

The USSR's anticoalition strategy has always combined diverse and seemingly contradictory components whose weight varies but whose simultaneous presence is a stable feature of the strategy. For example, Brezhnev presided over continuous competition between economic conservatives who favored closer commercial ties with Eastern Europe, on the one hand, and economic reformers who favored closer commercial ties with the United States and/or

Western Europe and Japan, on the other. In the early 1970s the former were ascendent; in the late 1970s and early 1980s the latter prevailed. Gorbachev, however, has called for closer CMEA-EEC ties *and* tighter integration of CMEA. Whereas Brezhnev seems to have viewed East-West and intra-CMEA trade as mutually debilitating, Gorbachev views them as mutually reinforcing. But unlike Brezhnev, Gorbachev is quick to emphasize the pitfalls of East-West trade.

> We must admit, after all, that we understood too late what kinds of traps are laid on the trading routes leading to the West. It has been discussed here what great losses have been suffered in Poland, and not by it alone. The very idea that it is simpler to buy from the capitalist market than to make it yourself has already done harm. Now we are decisively uprooting such inclinations in our country. It is not, of course, a matter of folding up economic links with the West. It is a different matter. It is a matter of making rational use of them, eliminating excesses and preventing dependence. Priority must be given, of course, to cooperative ties with the fraternal countries and to accelerating the process of socialist economic integration.[26]

Also, present-day Soviet analysts draw a sharp distinction between the economic-political and military-political integration of Western Europe. The former can be but is not necessarily desirable; the latter is always highly undesirable.[27] For example, a *Pravda* editorial observed:

> It would be useful to establish more businesslike relations between the EEC and CMEA. The CMEA countries' constructive initiative in this direction has apparently had a favorable reception. It is important that it produce real results. To the extent, moreover, that the EEC countries are a "political unit," the CMEA countries are prepared to seek common ground with them—in various forms, including parliamentary ties—on specific international problems, too.[28]

Yet Yurii Zhukov, a leading *Pravda* commentator, subsequently contended that the economic integration of Western Europe may lead to its military-political integration and that "the forces which advocate the creation of a Western military-political union, in which the FRG would be the pivot, are sticking to their guns." To be sure, Zhukov maintained that efforts to integrate the highly industrialized European nations more rapidly than the others and to "reform" EEC institutions were meeting sustained internal resistance, particularly over "demands to extend the powers of the community's executive organ, the Brussels Commission, and the functions of the European Parliament." But Zhukov concluded: "All the same, the maneuverings of those who advocate turning the EEC into a military-political union should not be underestimated. This design takes a long-term view. And there can be no doubt that its authors, chiefly those in Bonn, will continue pushing it forward, if only in the form of a 'two-speed Europe,' starting with the strengthening of military-political cooperation between France and the FRG."[29]

The Gorbachev leadership affirms that stronger East-West political, economic, and scientific-technological cooperation can, but does not necessarily, reduce military confrontation. West European nations are thought to have furthered detente considerably in the political, economic, and scientific-technological spheres but only marginally in the military sphere. Progress in all four spheres is deemed essential to European security, which the United States is allegedly jeopardizing. Fyodor Burlatskii elaborated on this theme in the following statement:

> The great, truly historic meaning, I would say, of the Helsinki conference of 1975 lies precisely in the fact that it was a major step along the path toward overcoming the division of Europe, along the path toward strengthening the security of the whole of Europe and the development of all-European cooperation. Looking back one can say that this process has been fruitful despite the truly colossal efforts made by its opponents to slow it down or to undermine its momentum. And here I would like to draw your attention to one very curious and even paradoxical thing: Most complicated of all in the way it has proceeded has been the process of movement toward military detente, and most fruitful of all has been the development of economic and scientific-technical cooperation.
>
> Well, in fact this paradox is easy to explain. The fact is security problems have largely been resolved for the countries of Western Europe by dictation from Washington, while they have decided problems of economic cooperation primarily by following their own interests.[30]

In a word, the Gorbachev administration is trying to forge links with key West European countries and collectively to take a more assertive stance vis-à-vis the United States on arms control and commercial issues.

Reassessing the Effectiveness of the Traditional Anticoalition Strategy

During the early 1970s, Soviet leaders held to a cautious but increasingly optimistic view of detente in Soviet-U.S. and especially Soviet-Western European relations. Optimism about Soviet-U.S. relations eroded in the late 1970s and declined precipitously in the early 1980s, but optimism about Soviet-West European relations was buoyed by the minor effects of the 1979 Afghan occupation and of the 1980–81 Polish crisis on East-West detente in Europe.[31]

Soviet frustration and consternation with the resurgence of U.S. military initiatives under the Reagan administration and with the ebbing assertiveness of

Western Europe in the early 1980s reflect a greater awareness of the cohesive tendencies of the Western Alliance and of the limits of Soviet influence over West European nations. These limits were becoming increasingly apparent to Soviet bloc and Western politicians and were being expanded by the mounting economic difficulties in the USSR and Western Europe as well as by the accelerated modernization of the U.S. and West European strategic forces. The Soviet leadership responded to these unexpectedly negative circumstances by terminating the INF talks in November 1983 and by focusing its anticoalition strategy on confrontation with the United States. A year later the Soviet leadership responded to the still worsening domestic and international conditions by agreeing to reopen East-West strategic arms negotiations and by focusing its anticoalition strategy on cooperation with the United States.[32]

When Gorbachev became general secretary in March 1985, he intensified the emphasis on U.S.-Soviet collaboration and sought closer ties with the major West European nations in the hope that they would influence the United States on key security and economic issues. He also sought closer ties with the smaller and less industrialized West European nations in the hope that they would influence the larger and more industrialized nations or play an intermediary role in East-West negotiations. Hence Gorbachev pursued a U.S.-centered policy while trying to establish viable fallback positions if his European peace offensive and U.S.-Soviet summit talks failed. Soviet strategy and tactics toward Western Europe became increasingly differentiated and more closely tailored to changing conditions in the USSR, United States, and Eastern and Western Europe. As Bovin put it:

> Of course, Britain, France, and West Germany are the first violins in the Western European orchestra, and we are dealing with them. But we are now stepping up our policy in regard to the small and medium-sized countries of Europe. Experience has shown that it is sometimes easier for small and neutral countries to feel out some kind of ground for compromise, especially during a period when mutual relations between us and the Americans, for instance, between East and West, are somewhat strained.[33]

In mid-1986 the increasingly open disagreements among the Western governments and within the U.S. government on vital security issues, especially the U.S. abandonment of SALT II ratification, gave the Soviet leadership renewed hopes of curbing or reducing U.S. influence in Western Europe. Also, the Gorbachev administration seemed less pessimistic about influencing the United States via its major allies, the FRG and Great Britain, and about gaining advantages for the Soviet bloc from divisive tendencies in the Western Alliance's policies and institutions. Furthermore, Gorbachev appeared certain that sizable security and economic benefits could accrue to the East and West from increased "realism" in the White House. Gorbachev and other

Soviet reformers anticipated that the seeds of such sober-mindedness had been planted at the first Reagan-Gorbachev summit in Geneva and might be harvested at a second summit. From the Soviet perspective the frustrations of dealing with a frequently truculent and indifferent and sometimes divided and disorganized U.S. administration heightened anticipation of the probable advantages of dealing with Reagan's successor and future Social Democratic Party (SPD) and Labour governments. However, the Soviet expectation of prompt U.S. congressional and public support and widespread European support for an arms control agreement presented by a conservative U.S. president and endorsed by conservative West European leaders, together with Gorbachev's eagerness to transfer scarce resources from military to civilian sectors to meet pressing economic difficulties, impelled the Politburo to seek short-term rather than long-term accomplishments.[34]

The West European reaction to the Reykjavik summit was distressing to the Soviet leadership and produced diverse Soviet explanations, which reflected a considerably intensified private and public debate about strategy and tactics vis-à-vis the United States and Western Europe. All Soviet analysts were concerned that Gorbachev's efforts to improve East-West relations had stimulated pronuclear sentiment in major West European governments. But some analysts stressed the continuity in Western Alliance policy, while others stressed the change. Some claimed that they had anticipated West European behavior, while others acknowledged that the West Europeans had surprised them.

Soviet ultraconservatives heaped blame on the United States for orchestrating West European responses. Military analyst Vladimir Bogachev declared: "Washington gave the European NATO countries the main role in undermining the Soviet-American accords on the issue of medium-range missiles that were reached in Reykjavik. As though responding to a command, some leaders of these countries, including Thatcher, Kohl, and Chirac, expressed 'alarm' at the possibility of a withdrawal of American medium-range nuclear missiles from Western Europe."[35]

Soviet conservatives especially blamed the United States, but they also accused West European governments of having once again sacrificed their national interests, of having feigned support of arms control prior to Reykjavik, and of having abruptly changed their policies on military detente after Reykjavik.[36] *Izvestiya* commentator Stanislav Kondrashev affirmed:

> Bonn, London, and Paris have ceased to support Ronald Reagan's "zero option," which envisaged the elimination of all Soviet and American medium-range missiles in Europe. This has not happened out of opposition to Washington but since the Soviet Union agreed to this option in Reykjavik—laying aside, incidentally, its own quite valid objections to the British and French missiles. Since the Soviet concession in Reykjavik the

U.S.'s Western European allies have hastily restructured their arguments in order to reject this concession. Now they are insisting that nuclear disarmament in the European theater would be dangerous for them because it would supposedly weaken the American nuclear shield and increase Soviet superiority in conventional weapons. It turns out that they prefer a Europe filled with nuclear weapons and are even prepared to reconcile themselves to the Soviet missiles as long as the American ones remain, as long as Europe does not become nonnuclear.

This kind of political and propaganda about-face in Western European capitals forces one to stop and think how serious they are in looking for ways to reduce tension. Well, here too, the picture is becoming clearer and more real.[37]

Soviet centrists seemed uncertain about whom to blame for West European resistance to the tentative Reykjavik agreements but were inclined to believe that conservative West European governments had acted independently and had successfully influenced the United States in certain negative ways. Consider the following exchange between *Novoe vremya* editor Viktor Tsoppi and international affairs correspondent Aleksandr Antsiferov:

Tsoppi: Let us recall that when Reagan, in conditions quite different from the present ones, put forward his zero option, NATO's Western European members applauded it tumultuously. But now, when an accord on scrapping both Soviet and U.S. medium-range missiles in Europe has been reached in principle in Reykjavik, these same people, these same statesmen are loudly indignant and, I would say, are even lamenting from fear. British Prime Minister Margaret Thatcher went to Washington a few days ago to persuade Reagan to renounce everything that Reykjavik has led to, and, judging by everything, she succeeded.
Antsiferov: What I don't understand is who was trying to persuade whom.
Tsoppi: Yes. And in Washington the following connection can be traced: In Washington, as they heed these revelations, they justify their own rejection of the Reykjavik accords, they justify it by the fact that, as noble people, they can't ignore the interests of their allies. That's how imperialist solidarity, or, if you like, mutual assistance operates.[38]

Soviet reformers reaffirmed the importance of constructing a "common European home," but they regretfully acknowledged that some West European leaders reacted to Reykjavik with a misguided and possibly cynical understanding of their nation's interests and that such leaders' behavior and attitudes could be difficult to change. Reformers also held out the hope that West European governments and citizens would pressure the United States to make arms control compromises and agreements, but they accused Reagan of sacrificing his allies' interests by not agreeing unilaterally to Gorbachev's "concessions" regarding the zero option and the French and British missiles. Politburo member and Foreign Minister Eduard Shevardnadze voiced such views about opportunities and dangers in a blunt, cajoling, and moralistic part of an otherwise temperate speech to CSCE representatives in Vienna:

> There is a lack of logic in the position of some European leaders on nuclear disarmament issues. When there finally arose the real possibility for clearing the continent of missiles they began to speak of the need to retain U.S. nuclear weapons in Europe and to defend their imagined privileges to nuclear status.
> What a lot of furious words erupted at one time concerning our stand on British and French nuclear arsenals. But now that a concession which is bold and to some extent even risky for us has magnanimously been made, we are offered a modern version of the comedy "Much Ado about Nothing." As if nothing had in fact happened, as if we had not taken such a serious and responsible step to accommodate our partners. There are even some cynics who are now saying that the NATO governments never seriously wanted this. They put up their argument about the British and French weapons because they were certain that the Soviet Union would never accept it. In other words, they were bluffing, engaging in demagogy.
> Now, instead of saying "We will join with you in time," they are all but declaring their nuclear weapons to be eternal.... It is a pity that some political leaders have proved to be unprepared to think in terms of a nuclear-free Europe.... This all-European home that we are building and in which all are equal will not become reliable and strong if deceit, half-truths, and disinformation are mixed into the mortar that holds it together. A shortage of trust must not be created because of false understanding of national prestige or election campaign concerns.[39]

Finally, Soviet ultrareformers stressed the differences between U.S. and West European interests as well as the willingness and ability of West European politicians and citizens to pursue their national and global interests after Reykjavik. These Soviet analysts anticipated renewed efforts to create a nuclear-free world by West European governments, opposition parties, and political movements as well as much closer Soviet and East European relations with Western Europe and more "realistic" Soviet-U.S. relations. International affairs commentator Nikolai Shishlin, taking a long-term view of evolving phases in West-West and East-West relations, optimistically concluded:

> It's quite a routine thing for the NATO countries, when developments reach an acute turning point, to demonstrate, to use a phrase, a united front. But we mustn't just look at the surface of these events. The united front is indeed a reality, but at the same time it is to some degree a myth, because the Western Europeans have different interests from the United States. If the United States pursues here, there, and everywhere an imperial policy, a policy of global confrontation with the Soviet Union, then the Western European partners of the United States may sacrifice Western European interests within the

framework of a global confrontation with the Soviet Union and the socialist world. These fears make themselves felt in the political action, the political steps, that are taken by the West....

Western European [leaders]... discover in talks with representatives of the socialist section of Europe that we have common interests, and these common interests are not limited to interests of preserving peace in this fragile European home of ours; they also include the need to develop peaceful, businesslike cooperation and a joint approach to pan-European problems.

So I think that while we can say there is a bloc discipline, we must nevertheless not put everything down to bloc discipline and see in the political picture of Europe today only this bloc discipline.... Now, after Reykjavik, not only has the door not been slammed in relations between the Soviet Union and the United States, but neither has a single door been slammed in the relations of the Soviet Union and the other socialist countries with the Western European countries.[40]

Thus, Soviet spokespersons had different views of contemporary East-West and West-West relations. But all contended that West European nations could be unduly subservient to U.S. interests and unmindful of their sovereign interests and that West European political and economic elites had the capabilities though sometimes lacked the will to resist U.S. hegemony. The cohesiveness Washington "subjectively" imposed on Western Europe was perceived in Moscow to be temporary because it intensified the "objective" contradictions in the interests of the United States, Western Europe, and Japan. Such cohesiveness was seen to be dangerous because it perpetuated confrontational methods of resolving East-West and North-South political, military, and economic problems. According to Gorbachev:

Washington is continuously calling on its allies not to waste their gunpowder on internecine strife. But how are the three centers of modern-day imperialism to share one roof if the Americans themselves, manipulating the dollar and the interest rates, are not loath to fatten their economy at the expense of Western Europe and Japan? Wherever the three imperialist centers manage to coordinate their positions, this is more often than not the effect of American pressure or outright dictation, and works in the interests and for the aims above all of the United States. This, in turn, sharpens, rather than blunts, the contradictions.[41]

Before Reykjavik, mounting Soviet exasperation with Western Europe was amply revealed in various Soviet forums.[42] After Reykjavik, top Soviet officials and analysts held different views about the likelihood of "realistic" West European influence on the United States. Valentin Falin, then candidate CPSU Central Committee member, Novosti press agency head, and former ambassador to West Germany, was considerably more pessimistic in an interview with West German editors of *Spiegel* than was Bovin in his interview for East German radio:

Spiegel: Now, after the unsuccessful exploratory talks in Reykjavik, is the starting signal being given for the development of a strategic defense system for the USSR?
Falin: That depends on what the Americans do. We will not expose ourselves to any danger.
Spiegel: But are you still hoping that the Americans' partners, in particular the Western Europeans, will bring pressure to bear on Washington?
Falin: We do not harbor such illusions.
Spiegel: However, the Europeans did exert some pressure on Washington, at least internally.
Falin: Certainly. The early-warning time for Europeans is so short that in effect they could expect a catastrophe at any minute.
Spiegel: So you expect Reagan to come under more pressure following Reykjavik?
Falin: We are realists and understand that neither Reagan nor we conduct negotiations under pressure. However, we of course assume that public reaction will also urge Reagan seriously to consider the problem.[43]

Bovin, responding to a question about the possibility of West European nations' playing a larger role in the pursuit of East-West arms control agreements, replied:

Yes, definitely, definitely. The wave of neoconservatism which has arisen not only in America, but also in Western Europe, is contributing to a certain degree to the fact that Europe is, in part, losing its own face. The tendencies of Atlanticism are fairly strong; the tendencies to hide behind the Americans are fairly strong; the tendencies to view the world through the eyes of the Americans are fairly strong. But, nonetheless,...there are political forces, movements, and parties in Western Europe, I am convinced, which advocate a more independent European policy, a more independent role for Europe in NATO...and in world policy generally.

Our position, Moscow's position—and we have never made a secret of it—consists of supporting Europe in this respect, of contributing toward Western Europe, and toward Europe as a whole, becoming a factor in world politics. No damage emerges from this, but it brings only benefits—both for Europe and for us, and for the Americans. We want good relations with America and Western Europe.[44]

With major arms control proposals emanating from the USSR and initially ignored or resisted by the United States, Soviet commentators saw "Atlantic unity" being undermined and "European unity" being enhanced. European unity was conceptualized (in descending order of importance) as acceptance of Soviet foreign policy initiatives and resolution of Soviet-West European conflicts generated since the late 1970s, closer ties between NATO and the Warsaw Pact and between the EEC and CMEA, and selectively closer ties among the West European nations themselves for purposes other than defense coordination and political and economic destabilization of Soviet bloc countries.

By renewing Soviet calls of the 1930s for "collective security" and a "united front," Soviet commentators not only implied that contemporary U.S. imperialism was analogous to Nazi aggression but that revanchist tendencies still lurked in Western Europe. The 1986 CPSU program proclaimed that "respect for the territorial-political realities which came about as a result of World War II is an inalienable condition for the stability of positive processes in this and other regions. The CPSU is resolutely opposed to attempts to revise these realities on any pretexts whatsoever and will rebuff any manifestations of revanchism."[45] Specifically, Soviet spokespersons periodically revived their warnings about United States-instigated FRG revanchism while maintaining and even expanding Soviet-FRG trade, credits, and scientific and technological cooperation. According to a *Pravda* editorial, "If you look back at the postwar period, it is indisputable that the rebirth of militarist tendencies in Western Europe was stimulated to a significant extent by forces outside the continent, mainly U.S. imperialist circles. The dangerous revival of revanchist forces in the FRG always takes place in precisely this atmosphere of the arms race. Encouraging revanchism runs counter to the interests of ensuring peace, detente, and cooperation on the continent; it is impermissible."[46]

Of much greater operational significance were Soviet claims that the Reagan administration was sacrificing European security in the pursuit of military superiority over the USSR and for the purpose of accelerating the East-West arms race. Boris Yeltsin, then a candidate Politburo member and probably the most frank and reformist top Soviet official, was quoted as follows in TASS, the official Soviet news agency:

> The United States wanted to turn Western Europe into its "dual hostage"—both nuclear and chemical, Yeltsin said. "As far as the Pershing II and long-range cruise missiles are concerned, the United States is clearly engaged in unfair play in a bid to divert a retaliatory strike from its own territory at the expense of the population of its allies, including the FRG.
> "This is a direct deception of the peoples of Western Europe. You and we cannot place the destinies of peace in the hands of American imperialism and the FRG government obediently following in its wake."[47]

Moreover, some Soviet commentators contended that the United States was conducting economic and psychological warfare against West European elites and populations in the hope of gaining fuller compliance with its militaristic preferences. But other Soviet observers maintained that the propaganda and intelligence agencies of the United States and its allies were working closely together to impede cooperation between the EEC and CMEA and between NATO and the Warsaw Pact as well as to undermine domestic and international support for the Soviet bloc's policies and social systems.[48] To counter such perceived security and subversive threats, Soviet policy toward the United States mixed demonstrations of the USSR's military power with affirmations that this power would be used only for defensive purposes. Soviet policy toward Western Europe mixed quiet and infrequent references to the USSR's military capability with loud and frequent affirmations about the USSR's conciliatoriness and flexibility in trying to enhance European security and world peace.

The Gorbachev administration has been less satisfied with the accomplishments of the traditional anticoalition strategy and more apprehensive about its potential than was the Brezhnev administration. Soviet conservatives especially would agree with a *Krasnaya zvezda* editorial's proclamation that "the socialist world is opposed by a strong and dangerous enemy—imperialism—which, in today's conditions poses a growing threat to mankind's very existence."[49] Soviet reformers especially would be cautiously optimistic about the Warsaw Pact's renewed call to the NATO countries "to display realism and responsibility in the aim of achieving accords taking into account the interests of both sides and all other states on the radical reduction of nuclear weapons and their subsequent elimination and on the prevention of an arms race in space."[50]

Yet virtually all Soviet analysts have based their strong convictions and reasoned arguments on the key premises that the USSR consistently pursues "greater security and stability in the world"[51] and poses no threat to its European neighbors. Soviet officials and commentators have found it exceedingly difficult to perceive or comprehend why the leaders of the Western Alliance have failed to accept these premises, other than to serve the selfish interests of their "military-industrial complexes." From the dominant U.S. viewpoint, the Soviet bloc's offensive force structure on the central European front and its military assistance to "progressive" Third World regimes and movements are highly disruptive forces in world politics and belie all Soviet claims of support for peaceful resolution of international conflicts. From the dominant Soviet viewpoint, U.S. expansion of the arms race into outer space and intensification of the arms race on the European continent are the most disruptive forces in world politics and belie all U.S. claims of support for peaceful change in East-West and North-South relations.

How could the USSR's anticoalition strategy more effectively enhance international security? Soviet reformers stress the growing power of "realistic," "sober-minded," and "cool-headed" leaders and citizens in the United States and Western Europe to resist the "war-mongering" and "anti-Soviet" impulses of their governments and politicians. These Western "peace-loving" forces are thought to be able to erode "conservative" and "neoconservative" public opinion on international affairs, thereby influencing conservative and neoconservative policymakers in the executive and legislative branches and eventually replacing them. Soviet reformers and Western realists decouple East-West economic and political-military issues and take "progressive" stands on North-South issues, opposing all multinational corporation profiteering but supporting or acquiescing in many "national liberation movements." According to Arbatov, one of the most informed though strident critics of the United States:

> A serious debate is now taking place in the West, and that includes political circles and the broad public in the United States, too. A struggle lies ahead, in the course of which the polarization will be stepped up between the mighty forces of the military-industrial complex, militarist forces, and extreme right-wing politicians on the one hand, and on the other those sober-minded political and public circles which understand U.S. national interests differently, or, to put it more simply, correctly.[52]

Soviet conservatives stress the "bourgeois class" bonds among the "imperialist" nations and the converging interests and growing political power of their military-industrial complexes. These complexes allegedly perpetuate the arms race and the cold war, preserve the Western Alliance's dominance of the world political economy, and further capitalist interests by exploiting the proletariat and peasantry of all nations and by suppressing "national liberation movements." Soviet and Western conservatives call for greater cohesiveness in their own polities and for greater vigilance against each other's machinations throughout the world. According to ultraconservative Major General A. Serebryannikov:

> The wave of mounting mass chauvinist psychosis, visceral anticommunism and anti-Sovietism, and frenzied war propaganda created by U.S. politicians is to a considerable extent carrying them toward the shore of strong-arm methods. The ideology underlying imperialism's desire to restore its former dominance now proclaims as its aim not simply "rolling back" communism, as was the case during the "cold war" period, but completely "eradicating" it and restoring global imperialist domination.
>
> In this sense, imperialism's "new crusade" is directed against all freedom-loving peoples and all liberated countries. It is essentially a policy of global terrorism relying on military force and directed against all progressive, freedom-loving, and peace-loving mankind. It is a concentrated expression of the class interests of world imperialism's most reactionary circles.[53]

Soviet reformers and conservatives argue that the United States is able to pressure its allies largely because of their common acceptance of "the myth of the Soviet threat." The nature of this perceived threat is described with sophistication and clarity by Bovin:

> There are hardly any politicians left who believe that the Soviet Union intends to seize and conquer Western Europe. The "Soviet threat" is now presented

more subtly and intelligently. Relying on its military-strategic supremacy, the Soviet Union—this is how they set forth our intentions—wants to bring Western Europe under its political control without war and turn it into a sphere of its predominant political influence. That is the end, and the means is to set Western Europe at odds with the United States and thereby break the U.S. security guarantees and leave Western Europe on its own with the Soviet Union.[54]

Soviet reformers are more optimistic than conservatives that efforts to dispel this "myth" will increase the West European nations' pursuit of their own and general European interests. Soviet conservatives are less optimistic than reformers that the power of imperialist military-industrial complexes will diminish in the foreseeable future, even under West European social-democratic governments. Also, reformers stress the independence and diversification of national capitalist economies and their competition with one another in the international economy. Conservatives stress the integration and militarization of national capitalist economies and their collaboration with one another in the international economy.

Soviet reformers and conservatives agree that the Western Alliance threatens the Soviet bloc. But they disagree about the nature, magnitude, and manageability of this threat. Reformers place greater weight than conservatives on commercial ties with Western Europe, Japan, and/or the United States. And conservatives place greater weight than reformers on commercial ties with CMEA. Also, reformers stress the advantages to the Soviet economy and to the USSR's anticoalition strategy of the improving and potentially excellent trade relations with Western Europe. And conservatives stress the disadvantages to the Soviet economy and to the USSR's anticoalition strategy of expanded East-West trade.[55]

Both reformers and conservatives criticize the political and economic stagnation of the last half of the Brezhnev period. But reformers view stagnation primarily as a structural problem that is considerably influenced by the world economy and that could be remedied in large part by domestic institutional reforms and by improved East-West relations. Conservatives view stagnation primarily as a leadership problem that is moderately influenced by the world economy and that could be remedied largely by perfecting domestic institutions and strengthening Soviet bloc ties.

Gorbachev has placed more and more emphasis on reformist structures and orientations in East-West and East-East relations. He sees close East-West commercial ties as a vital part of the USSR's response to domestic economic problems, especially declining productivity increases and low-quality manufactured goods, if the USSR can avoid international economic and political disparities. Also, he considers much closer integration of the CMEA economies—better multilateral planning combined with more market incentives—to be essential to Soviet economic modernization. Gorbachev's initial changes in Soviet foreign trade decision-making, such as doubling the number of foreign trade organizations, enabling them to deal directly with West and East European suppliers and buyers, and encouraging Soviet-Western and intra-CMEA joint ventures, constituted a victory for market-oriented political leaders, production executives, and economists throughout the Soviet bloc. Likewise, Gorbachev's contentions that export-oriented production and international competition would upgrade the quality of Soviet manufactured goods and the technological level of manufacturing processes loudly echoed reformist arguments of the 1960s and 1970s.

But Gorbachev, with an acuity and a sense of urgency that place him in the forefront of economic and strategic modernizers, understands that the East-West arms race has diminished Soviet security. Scientific-technological competitiveness in the civilian economy has been the primary casualty to date, and scientific-technological competitiveness in the military economy is the likely casualty in the not-too-distant future. To forestall these alarming trends, the Gorbachev administration has begun to adjust key concepts, elements, and instruments of Soviet foreign policy.

Political, Economic, and Military Responses to SDI and INF

Gorbachev and his fellow reformers perceived the heightened East-West tensions of the early and mid-1980s, especially the U.S.-initiated SDI program and installation of INF systems in Western Europe, to be economic as much as security challenges. Soviet research and development had long concentrated on defense industries, and spin-offs from the military to the civilian sector had been few. These trends would continue or even accelerate if the Warsaw Pact countries had to counter NATO's apparently enhanced first-strike nuclear capability and information technologies. The West's sophisticated computers and telecommunications would further weaken the Soviet bloc's economic competitiveness and would have destabilizing military applications in outer space as well as in strategic and conventional forces.

The Reagan administration's strategic and conventional arms buildup, SDI research, and INF deployment made the Gorbachev administration reassess Brezhnev's military assumptions and emphasize the multifaceted causes, character, and consequences of security. The present Soviet leaders see economic progress as more dependent on arms control than ever before. SDI jeopardizes the USSR's economic growth and productivity by draining research and development resources sorely needed to modernize information technologies, to upgrade an aging capital stock, and to produce high-quality industrial and consumer goods. Also, SDI reduces the competitiveness of Soviet bloc conventional forces and limits the effectiveness of strategic arms agreements. Furthermore, Soviet officials consider INF and SDI to be integral parts of a rejuvenated U.S. effort to dominate the Western Alliance. They think SDI is especially pernicious because it shifts West European and Japanese research and development from civilian to military sectors, reduces West European and Japanese motivation and capability to expand economic ties with the USSR, and integrates under U.S. hegemony the modernized defense establishments of Western Europe and Japan.

Gorbachev's Politburo viewed good relations with the United States and Western Europe as complementary. To be sure, top CPSU officials disagreed about the likelihood and desirability of various kinds of Soviet-American cooperation. But the Soviet leadership has revived constructive communication with the United States, including five summit meetings between Gorbachev and Reagan, periodically serious arms control negotiations at various levels, and unprecedented people-to-people exchanges, both on television and in person. The USSR also has revitalized bilateral ties with many large and small West European countries, indicated a willingness to deal with the EEC as a political unit, and begun unilaterally to demobilize half a million troops at home and in Eastern Europe and Mongolia. Gorbachev has hoped to influence the conservative U.S. and West European governments directly, which may in turn influence U.S. officials. Gorbachev would have welcomed social democratic governments in Western Europe and a liberal administration in the United States, but he did not expect them to come to power soon. Indeed, after Gorbachev and Reagan agreed in December 1987 to withdraw INF weapons from Eastern and Western Europe, the Soviet leadership probably quietly rooted for a George Bush victory over Michael Dukakis in the hope of building on a stable foundation with experienced and familiar personnel. "Continuity" was the watchword in Soviet public diplomacy.

Gorbachev has sought mutually beneficial military, economic, and scientific-technological accords with existing Western governments and multinational institutions, especially the United Nations, NATO, EEC, Western European Union, International Monetary Fund, General Agreement on Tariffs and Trade, World Bank, and multinational corporations. The traditional Soviet "wedges"—propaganda, inducements, and threats used to separate Western Europe and the United States, individual West European countries, West European governments and their citizens, and political and economic issues—have not been unilaterally discarded. Rather, Gorbachev is using them more selectively in support of national defense and domestic economic modernization. Soviet leaders are trying to work with current Western institutions and conservative officials. Efforts to weaken these institutions and remove such officials through appeals to Western public opinion proved counterproductive in the late 1970s and early 1980s, especially in West Germany. Also, new or newly acknowledged internal and external conditions had created pressures for arms limitation and foreign trade, to which Gorbachev responded with a purposefulness and a gift for media relations unprecedented in a CPSU general secretary.

The Gorbachev leadership has become increasingly sensitive to the interconnections between the economic and security elements of its anticoalition strategy. SDI research and possible deployment underscored the need to tailor the USSR's anticoalition strategy to a multifaceted new threat from the West and to an evolving Soviet socioeconomic and scientific-technological modernization strategy.

Soviet spokespersons rarely distinguish between the increasing West European support for SDI research and the continuing West European coolness to SDI deployment. The absence of this distinction is a feature of Soviet declaratory policy but probably not of actual policy. Soviet leaders fear that West European politicians, especially conservatives, are attracted to SDI as a means of keeping pace with U.S. scientific and technological advances and of coping with unemployment, inflation, declining productivity, and other domestic economic problems. Soviet analysts contend that West European industrialists and high technology firms are avariciously pursuing SDI contracts, with such deleterious long-term consequences as siphoning off skilled personnel, capital, and equipment from Western Europe to the United States and heightening European political, economic, and scientific-technological dependence on the United States. M. Kalachev poses and answers an important question:

> Why, then, is Western Europe as a whole taking part in the "space race"? There are several reasons. First, fear of showing allied disloyalty to the United States. Second, Washington pressure. Third, SDI champions in Western Europe ascribe their upsurge to a desire not to lag behind in the high technology of the future, which allegedly, only "star wars" can give to a fading Europe.
>
> However, the main incentive for taking part in this dangerous program is the ardent desire of Western European monopolies to obtain a slice of the immense profits which SDI promises. It is this greed and pursuit of promised millions which spur private capital and the governments of Western Europe that protect its interests to step onto the "star wars" running track.[56]

Some Soviet commentators, well aware of West European and Japanese hopes to reap a commercial bonanza from the militarization of space, argue that SDI contracts exploit rather than strengthen the United States's allies. Leonid Ponomarev asserts:

> The "economic benefits" promised by Washington, by ideas and technologies which [may eventually] stem from the "star wars" program, will surely

benefit primarily the American monopolies. Western European and other companions will undoubtedly find themselves in an inferior position, since the "brains" brought up by Washington will leave their "products" in the United States. It is natural that the "brain drain" will inevitably weaken the economies of those countries in general, for they will be unable to compete with the United States on a par. Americans do not bother to conceal that they are hunting for everything new, even if that damages the interests of their friends.[57]

A duller wedge than Ponomarev's is hard to imagine.

In criticizing West European eagerness to conduct SDI research, the Soviet leaders found themselves in a dilemma. On the one hand, they understood that SDI was creating fissures in the Western Alliance by drawing Bonn and Washington closer together. On the other hand, Paris responded by trying to enhance Franco-German security ties. Neither alternative was palatable to the USSR. Hence, SDI made it more difficult to keep wedges between the United States and West Germany and between West Germany and France.

Yet Gorbachev and his colleagues linked the new specter of SDI with the old specter of a nuclear-armed West Germany. Also, they tried to persuade West European nations to define and pursue their interests more independently of the United States and of each other. And, whenever possible, Soviet spokespersons cited Western supporters of the USSR's policies. As V. Gusenkov affirmed:

> A ... widespread "argument" of the adherents of a "European defense" in France—there are many of them in the government and the opposition—is reduced to this assertion: Arresting the FRG's slide toward Washington can only be done by strengthening the French–West German military alliance. The thinking is that a convergence with the FRG in military-political questions would prompt it to support France's policies in Western Europe. This is, of course, a delusion. The Paris newspaper *Le Monde* expresses itself quite definitely: The Bonn coalition will not make any decision that could create the impression that it is dissociating the FRG from the United States. The Rhine clearly does not intend to sacrifice its close ties with the United States for the sake of strategic cooperation with France.
>
> Farsighted politicians in France are increasingly asking themselves the question: Will not the current attempts to draw the FRG into the orbit of military-political cooperation end up hurting French interests? Having enormous economic potential, the FRG is also becoming a leading military force in Western Europe....
>
> Bonn's participation in the "star wars" program opens the way to obtaining even more developed military technology. The French newspaper *Libération*, which is close to the government, draws attention to precisely this aspect: Will not Bonn, in circumvention of existing prohibitions, get access to nuclear weapons? A reasonable question.[58]

Also, Soviet leaders apparently assume that West European nations become more independent of the United States under peaceful and routine conditions and significantly more dependent on the United States under dangerous and crisis conditions. Indeed, West European security policies were becoming increasingly independent of the United States and of one another until the early 1980s. These trends were slowed by INF deployment and stalled by SDI. Hence top Soviet officials thought that the United States was augmenting its influence in Western Europe by producing East-West crises—real (such as SDI) and imagined (for example, the SS-20 "threat")—and that the USSR could augment its influence in Western Europe by avoiding such crises (for example, by not sending additional Soviet troops into Poland). Such views about East-West crises and the cohesiveness of the Western Alliance have been articulated by Falin and Arbatov:

> *Falin*: In conditions of detente and normal relations the weight of states, large ones like England and smaller ones, normally grows. The countries gain additional possibilities for self-expression, for representing and promoting their interests. In this sense the weight of the military factor created or not created by the superpowers is leveled out. That is a very important point.
> *Arbatov*: This situation of tension is a kind of emergency in which no one has the right to act according to his wishes.[59]

Soviet foreign policy under Gorbachev was also grounded on the assumption that West European nations, in order to maintain their sovereignty in East-West relations and to improve North-South relations, must join with the USSR and Eastern Europe to resist U.S. "state terrorism" (for example, the U.S. air strike in Libya, support for the government of South Africa) and "neoglobalism" (for example, U.S. anticommunist military and economic activities, especially in Third World countries, such as Nicaragua, Grenada, Angola, and Afghanistan). According to Bovin:

> The essence of the strategy of neoglobalism can be described as an attempt to halt and reverse social progress inasmuch as it is incompatible with U.S. "vital interests," with the preservation of the global U.S. military and political presence, and the U.S. claims to world leadership. Relying on its tremendous resources and on its gigantic war machine, in other words, relying on strength, the United States is challenging world history, clearly in the hope of "replaying" it, replacing the onward march of history with a perpetual cycle of capitalist forms of societal life.[60]

Gorbachev's foreign policy was further grounded on the assumption that American "state terrorism" and "neoglobalism" are greatly aided by the "new stage" of the worldwide scientific-technological revolution. Of particular importance are microelectronic applications spurring "the transnationalization of the [capitalist] military-industrial complex," "the transnationalization of [capitalist] politics and ideology," and "the transnationalization of financial capital." SDI was perceived as a serious threat because it would pioneer technological breakthroughs in space weaponry and would solidify military research and development linkages among the Western allies. Capitalist governments would thereby have an increasingly dangerous incentive and capability to coordinate their military, economic, political, and ideological strategy, to establish a united front against socialist states and national liberation movements, and to achieve "social revenge on a global scale."[61]

In fact, SDI was a threat to the USSR's most important short- and long-term aims vis-à-vis Western Europe and Japan. According to Robbin Laird, the current Soviet leadership sees three primary dangers emanating from the West and pursues seven critical objectives regarding Western Europe. SDI exacerbates all three dangers and at least five objectives. SDI heightens Soviet fears "that the West Germans will become more nationalistic; that the European NATO states will succeed, however imperfectly, in formulating a common military policy; and that the United States will nurture both developments while at the same time maintaining a formidable military presence in the region."[62] Furthermore, SDI greatly impedes the following Soviet objectives:

- To encourage U.S. isolationism and West European perceptions of the United States as an offshore power
- To discourage the security efforts of individual countries in Western Europe
- To exacerbate existing tensions in intra-European bilateral relations, especially Franco-German relations, while encouraging its own bilateral links with them
- To minimize Western Europe's global power and to encourage regional isolationism
- To discourage West European military integration, especially the development of West European nuclear capabilities[63]

SDI research—to say nothing of deployment—has had a significant impact on Gorbachev's domestic and international priorities. To be sure, Soviet concern about SDI was greater in the mid-1980s than in the late 1980s. But SDI has been an important variable in Soviet deliberations about whether to alter the traditional anticoalition strategy or merely to adjust it to new conditions.

Conclusion

The USSR's traditional anticoalition strategy toward the West has been modified but has not changed dramatically under Gorbachev. This strategy was generally effective and efficient throughout the post-Stalin period:

effective in helping to achieve the USSR's basic security, economic, and hegemonistic goals and efficient in using limited resources to influence the policies and structures of the Western nations, collectively and individually. The Soviet anticoalition strategy has always combined obstructive and constructive elements. Khrushchev's periodic emphases on the former produced some remarkable failures (for example, the Berlin crises), and Brezhnev's periodic emphases on the latter produced some impressive successes (for example, political and economic detente in Europe). Gorbachev has placed a sustained emphasis on the cooperative elements of peaceful coexistence and has coproduced an INF agreement, gained enormous popularity throughout the world, and reconceptualized the aims and methods of international politics—most notably in his 1988 address to the United Nations. There seem to be mounting domestic and international pressures on Gorbachev to revamp the USSR's traditional anticoalition strategy but many constraints on doing so.

Foremost among these constraints is the basic nature of Soviet foreign policy and policymaking. As Franklyn Griffiths thoughtfully affirms:

> Soviet actions are not likely to be well understood when they are viewed in terms of unilinear, internally consistent policy whose direction is deliberately changed by the leadership as circumstance requires.... Soviet conduct is better regarded as internally contradictory, consisting of a series of persistent tendencies whose relative strength alters in response to international and domestic situational variables.... Soviet policy-makers are best seen as "unconditional thinkers" who produce a mobile consensus as together they vary the correlation of regime tendencies in processing a continuous sequence of issues.... The result is a policy that at once displays remarkable continuity and a propensity to oscillate as tendencies are combined and recombined to produce an array of responses.[64]

These conclusions emerged from an analysis of different Soviet images of the United States prior to Gorbachev, but they are valuable to an understanding of Soviet policy toward the Western Alliance in the Gorbachev period. For example, shifts in the cohesiveness of NATO and the EEC are increasingly likely to produce incremental and timely shifts in Soviet policy toward individual West European nations.

Second, the competing priorities or tensions in the vaunted new political thinking constitute an important but diminishing constraint. In an early exposition of the new political thinking, Central Committee member and *Pravda* Editor-in-Chief Viktor Afanasev juxtaposed such concepts as interdependence, mutual security, and the "unthinkability" of nuclear war, on the one hand, with "class enemies," "uncompromising ideological struggle," and "just—defensive and liberation—wars," on the other.[65] By 1988, however, more and more Soviet leaders excluded class struggle from their interpretation of peaceful coexistence. Yegor Ligachev's effort to reverse this trend probably cost him his ideological and foreign policy portfolios and helped Aleksandr Yakovlev become head of the powerful new Foreign Affairs Commission. But Valentin Falin, surely not a radical reformer, succeeded Anatolii Dobrynin as head of the International Department. Hence the Gorbachev administration's worldview has contained some innovative ideas as well as ideas advocated by Khrushchev and Brezhnev and even by Lenin and Stalin. Although the relative importance of these ideas is changing, the competition among them continues to limit the parameters of change.

Third, leading Soviet analysts view divisions in Western political thinking as a permanent feature of the increasingly multipolar international system and of the increasingly differentiated capitalist societies. These divisions constrain would-be reformers of the USSR's traditional anticoalition strategy because "realistic Western leaders will come to power sooner or later and Soviet patience and restraint will eventually be rewarded. Yevgenii Primakov, head of the Institute of World Economies and International Relations, and Gennadii Gerasimov, head of the Ministry of Foreign Affairs' Information Department, observed:

> *Primakov*: A line should be drawn between the neoconservative concept [of international politics] held by Reagan and the conservative concept. We cannot say that everybody in Washington is a neoconservative, i.e., a person who believes that all questions should be resolved through strength. Furthermore, it does not apply to Western European leaders. The conservatives are now clearly divided into so-called idealist conservatives, or more accurately, ideological conservatives, and pragmatists. This division may become even wider. Our foreign policy strategy takes into account this division and the emergence in the future of more pragmatic leaders.
>
> Ideological conservatives see as their goal the need to end socialism and subordinate it to their political actions. Pragmatists, while they oppose socialism, do not aim to end it through use of force. Social reformists can be our partners in campaigning for peace, ending tensions, banning nuclear arms and other weapons of mass destruction. We are prepared to cooperate with them.
>
> *Gerasimov*: The main thing is to see how people understand differences between socialism and capitalism. Let's speak of game theory. If one takes a view of the conflict between socialism and capitalism as being expressly antagonistic, as being a zero-sum game, if one sees the conflict exclusively as being black and white, as a struggle between good and evil, one would arrive right at the neoconservative position.
>
> If the differences between socialism and capitalism are viewed as a plural game, as a conflict where the opponents have some common interests, like playing according to the rules and keeping peace, then we can also put under this umbrella the concept of peaceful coexistence and the numerous

political movements, since their proclaimed goals are only reachable in conditions of peace.[66]

The policy implications of cleavages among and within Western governments and between Western governments and their citizens are simple in theory but difficult in practice: help "realistic" Western officials and programs but hinder conservative officials and programs. And, although Western ecological and women's groups stress protest more than a "constructive approach," Falin advocates taking "an attentive and objective look at all new public movements, especially those rooted in a broad social base."[67]

Fourth, Soviet leaders take a long-range and positive view of shifts in Western public opinion and of negotiations with even the most conservative Western leaders. These perspectives also constrain inclinations to change the traditional anticoalition strategy. Falin, in an interview with West German journalists immediately after the Reykjavik summit, commented:

> *Spiegel*: Time is passing, at least in foreign policy.
> *Falin*: Under Gorbachev, the Soviet Union's international situation has improved. Confidence in our country has grown. People show more interest in our proposals and initiatives. The most striking example was the moratorium on nuclear tests, and I am sure that the proposals we submitted here [at Reykjavik] will meet with international support—the reduction of long-range weapons by half, the elimination of a large number of intermediate-range weapons, and the renunciation of SDI.
> What was done in Reykjavik could be called a long-term investment which does not immediately yield a dividend, but pays nonetheless. Reykjavik has not been an unnecessary expense, either internally or externally....
> *Spiegel*: What impression did Gorbachev gain of Ronald Reagan during the summit meeting?
> *Falin*: Gorbachev has said that it was the first time for Reagan to negotiate in a really businesslike way. Reagan has shown that he can negotiate constructively if he wants to. We have concluded from that fact that it is also possible to cooperate with the President.[68]

Note that Primakov categorized Reagan as a neoconservative, not a conservative, and that Falin implied that Reagan is a neoconservative with pragmatic leanings. Neither Primakov nor Falin hinted that substantial alterations in the traditional anticoalition strategy were desirable, possible, or imminent.

In summary, the current Soviet leadership wants neither a strong U.S. presence in Western Europe nor a powerful and independent Western Europe that is united politically, militarily, and economically. Gorbachev and his colleagues probably want the United States to remain in Western Europe but in a weakened condition. Soviet observers consider fissures in NATO and the EEC to be enduring components of West-West relations because most West

Europeans are seen as nationalistic or merely tolerant of U.S. political, military, economic, and cultural influence on the European continent. Soviet analysts also think that the periodically widening splits in the Western Alliance can help the USSR establish and strengthen ties with large and small West European countries. Gorbachev does not claim to have created or first recognized such possibilities, but he is more determined than his predecessors to fulfill this potential.

Important sources of Gorbachev's determination have been SDI, INF, stagnant Soviet economic growth and productivity, and fresh East European efforts to develop bonds with Western Europe. Also, Gorbachev has had considerable confidence in the short-term efficacy of his conciliatory image among the leaders and citizens of Western countries. He has had confidence as well in the long-term benefits of supporting Western "realists" and expanding their numbers rather than confronting or capitulating to "conservatives" and "neoconservatives." Gorbachev's diplomacy focuses on foreign elites and organized counterelites, but he and his spokespersons have used the Soviet and Western media to appeal quite successfully to world opinion. The Gorbachev administration has thereby utilized many direct and indirect means to influence Western policies and to help bring into power Western governments less conservative in their policies toward the USSR. In turn, Gorbachev has resisted Western efforts to influence the USSR directly or indirectly through its East European allies. He has tried, with little initial success, to revitalize CMEA and the Warsaw Pact by offering more leeway and inducements rather than directives and sanctions. As A. Ross Johnson observed: "East-West European ties constrain Soviet policy in Western Europe, limiting the extent and duration of sharp departures in the direction of wooing or threatening Western Europe, and at the same time dampening efforts to discipline Eastern Europe."[69]

In addition, Gorbachev thinks that a cohesive Western Alliance can further the USSR's security interests under various circumstances. Pressured by domestic disagreements, disappointments, and debacles, Gorbachev might have concluded that, regardless of the odds, he must try to persuade Reagan that arms limitation accords benefit both the East and the West. Some of Gorbachev's Politburo colleagues probably viewed the first and second Soviet-American summits as premature and ill-advised. But later they could much more easily fault the general secretary for not negotiating with the popular, image-conscious, and erratic president of the United States than for achieving meager results. In an especially confident and calculating mood, Gorbachev might have concluded that the USSR could gain from two different developments. On the one hand, if Reagan's considerable power in the Western Alliance were declining, the president might try to restore and legitimize his power by withdrawing Euromissiles and modifying or delaying SDI. On the other hand, if Reagan's power were increasing and he thought that an INF and/or SDI agreement benefited the West as well as the East, his preeminence in the Western Alliance would help overcome or override his allies' anxieties about their vulnerability to Warsaw Pact conventional forces. Either outcome would assist the restructuring of Soviet society and would provide greater security for the Soviet bloc.

More effective, informed, and integrated Soviet domestic and international policies have resulted from the new personnel and the innovative structures, style, and modes of thinking that Gorbachev has introduced and championed. But the incoming generation of Soviet leaders has not jettisoned the traditional anticoalition strategy, which is adaptable to different phases of East-West relations, takes a long-term view of international change, and has served the USSR well in dealing with Western conservative, liberal, and socialist governments. The present Soviet leadership has energized the traditional anticoalition strategy while adjusting it to an emerging strategy of socio-economic reform at home, a worldwide revolution in telecommunications and computers, and dilatory but less and less disquieting responses from the West. Only the future will tell whether these Soviet adjustments to new conditions will lead to significant modifications of the traditional anticoalition strategy. Gorbachev concluded his *Time* interview with a question that he himself probably will continue to ponder and other Soviet leaders will surely debate: "If we in the Soviet Union are setting ourselves such truly grandiose plans in the domestic sphere, then what are the external conditions that we need to be able to fulfill those domestic plans?"[70]

Notes

1. Research for this chapter was completed in January 1989.
2. For a comprehensive anthology of policy-relevant essays on Soviet international behavior, see Robbin F. Laird and Erik P. Hoffmann, eds., *Soviet Foreign Policy in a Changing World* (Hawthrone, NY: Aldine Publishing Co., 1986), especially Parts IV and V.
3. See, for example, Richard Herrmann, *Perceptions and Behavior in Soviet Foreign Policy* (Pittsburgh: University of Pittsburgh Press, 1986), especially pp. 3–49, 180–199.
4. See, for example, Alexander George, *Propaganda Analysis* (Evanston, IL: Row, Peterson & Co., 1959), especially pp. 13–57.
5. All quoted statements by Soviet officials, unless otherwise noted, are from primary source materials translated and dated in Foreign Broadcast Information Service, Daily Report: Soviet Union (FBIS). For example, these statements are by Mikhail Gorbachev, October 15, 1986, pp. DD9–DD10; Georgii Arbatov, November 14, 1986, p. AA9; and Valentin Falin, August 28, 1986, p. AA8.
6. Quoted in *RL* (New York: RFE/RL), No. 399/86, p. 4; also in *FBIS*, October 9, 1986, pp. R1–R3.
7. *FBIS*, November 12, 1986, p. CC3.
8. *FBIS*, March 10, 1986, p. O2.

9 *FBIS*, April 24, 1986, pp. AA10–AA11. See also March 28, 1985, p. CC7; July 1, 1985, p. CC15; February 19, 1986, Part Three, pp. O3–O4; March 12, 1986, Part One, p. O2, Part Two, p. O3, and Part Three, p. O1. Note the terseness and ambiguities in the last two key documents—Gorbachev's report to the 27th Party Congress and the Basic Guidelines for Social and Economic Development.
10 *FBIS*, CPSU program, Part III, March 10, 1986, p. O4.
11 *FBIS*, March 28, 1986, p. O31.
12 *FBIS*, July 25, 1986, p. R7. See also October 10, 1985, pp. G1–G2.
13 *FBIS*, September 25, 1985, p. G2; July 1, 1986, p. AA10.
14 *FBIS*, June 24, 1985, p. BB3. See also September 25, 1985, p. G2; October 21, 1985, p. CC4; July 26, 1985, p. BB1; July 30, 1986, p. AA5.
15 This Soviet perspective is important and deep seated but is rarely articulated. A Western analyst with a particular interest in and evidence regarding the perspective is Ronald Asmus. See, for example, *RFE* (New York: RFE/RL), No. 130/85, 144/85, and 35/87.
16 *FBIS*, July 1, 1986, p. F6.
17 *FBIS*, October 2, 1985, p. G24.
18 *FBIS*, January 17, 1986,. p. CC6, See also October 2, 1985, p. G7; October 10, 1985, p. G2; October 11, 1985, pp. G1–G2; February 21, 1986, pp. AA10–AA12; June 11, 1986, p. CC2.
19 For Soviet use of the word *enemy*, see *FBIS*, July 18, 1985, p. CC15; March 25, 1986, p. BB1; August 15, 1986, p. CC1; and December 17, 1986, p. CC3.
20 In Laird and Hoffmann, eds., *Soviet Foreign Policy*, pp. 443–445.
21 John van Oudenaren, *Soviet Policy toward Western Europe: Objectives, Instruments, Results* (Santa Monica, CA: RAND Corporation, Feburary 1986), p. 4.
22 For evidence regarding these four assumptions, see, for example, *FBIS*, September 5, 1985, p. AA3; November 1, 1985, pp. G7–G8; March 20, 1986, pp. G4, G7; May 29, 1986, p. G3. See also June 3, 1985, p. G2; August 28, 1985, pp. CC2–CC4; January 15, 1986, pp. CC5–CC8; July 28, 1986, pp. G2–G4.
23 The distinction between minimal and maximal goals is an important part of the Leninist tradition, but present-day Soviet officials and commentators do not articulate the priorities of the USSR's anticoalition strategy. The promise and pitfalls of various bilateral ties are often noted, however.
24 *FBIS*, October 4, 1985, p. G10.
25 *FBIS*, July 17, 1986, p. CC17. See also July 10, 1985, p. A3; February 4, 1986, pp. AA2–AA3; July 1, 1986, p. F6; July 22, 1986, p. CC9; July 24, 1986, p. CC2.
26 *FBIS*, July 1, 1986, p. F5.
27 *FBIS*, November 13, 1985, pp. AA4–AA5. See also August 2, 1985, p. G1; July 16, 1986, pp. CC5–CC6.
28 *FBIS*, November 13, 1985, pp. AA4–AA5.
29 *FBIS*, November 1, 1985, pp. G7–G8.
30 *FBIS*, August 6, 1985, p. G1. See also April 23, 1985, p. G5; March 20, 1986, pp. G3–G7.
31 See Erik P. Hoffmann and Robbin F. Laird, *The Politics of Economic Modernization in the Soviet Union* (Ithaca, NY: Cornell University Press, 1985); and Erik P. Hoffmann and Robbin F. Laird, *"The Scientific-Technological Revolution" and Soviet Foreign Policy* (Elmsford, NY: Pergamon Press, 1982).
32 The perspectives cited are inferred from the USSR's actions in the period, especially its decisions to end INF talks in November 1983 and to resume broader strategic arms negotiations in November 1984.
33 *FBIS*, July 29, 1986, p. CC5.
34 *FBIS*, June 9, 1986, pp. A1–A2. See also May 28, 1985, p. G5; June 9, 1986, pp. CC5–CC6; July 1, 1986, p. A8; July 24, 1986, pp. CC1–CC2.

35 *FBIS*, December 9, 1986, p. AA5.
36 See, for example, Vladimir Chernyshev's commentary, *FBIS*, November 25, 1986, p. AA3.
37 *FBIS*, November 4, 1986, p. AA7.
38 *FBIS*, November 24, 1986, p. AA7.
39 *FBIS*, November 7, 1986, pp. CC2–CC3.
40 *FBIS*, November 4, 1986, pp. CC1–CC12.
41 *FBIS*, March 28, 1986, p. O6.
42 See, for example, Spartak Belov and Vadim Muchkin, *FBIS*, June 17, 1986, p. CC4, and June 10, 1986, pp. AA8–AA9, respectively.
43 *FBIS*, October 24, 1986, p. A11.
44 *FBIS*, November 6, 1986, p. AA4.
45 *FBIS*, March 10, 1986, p. O4.
46 *FBIS*, November 13, 1985, pp. AA3–AA4. See also April 23, 1985, pp. G2–G3.
47 *FBIS*, May 5, 1986, p. G3. See also July 12, 1985, p. G2; September 9, 1985, p. CC7; February 13, 1986, p. AA6; June 17, 1986, p. CC4.
48 See, for example, V. Gan, *FBIS*, September 4, 1985, p. CC5. See also April 23, 1985, p. G1; July 12, 1985, p. A2; July 25, 1985, p. AA2; March 25, 1986, pp. BB1–BB2.
49 *FBIS*, March 25, 1986, pp. BB1–BB2
50 *FBIS*, March 25, 1986, p. CC3. See also July 17, 1985, p. A8; April 9, 1986, pp. AA1–AA2.
51 *FBIS*, August 28, 1986, p. AA8.
52 *FBIS*, March 12, 1986, p. AA3. See also December 2, 1985, pp. CC7–CC9.
53 *FBIS*, June 13, 1985, p. CC3. See also March 25, 1986, p. BB1.
54 *FBIS*, September 25, 1985, p. G1. See also July 15, 1986, p. CC2.
55 See Hoffmann and Laird, *The Politics of Economic Modernization*, and Hoffmann and Laird, "*The Scientific-Technological Revolution.*" See also Philip Hanson, "On the Limitations of the Soviet Economic Debate," CREES Discussion Paper GS 2, University of Birmingham, 1985.
56 *FBIS*, March 18, 1986, p. G4. See also August 5, 1985, p. G1; December 17, 1985, pp. CC2–CC4.
57 *FBIS*, August 8, 1985, p. AA2. See also March 19, 1986, p. AA7; April 10, 1986, p. G6.
58 *FBIS*, February 20, 1986, p. G4. See also March 28, 1986, pp. AA3–AA6; April 2, 1986, pp. AA7–AA10; July 28, 1986, p. G4; August 15, 1986, pp. AA4–AA9.
59 *FBIS*, July 17, 1986, pp. CC17–CC18.
60 *FBIS*, March 11, 1986, p. A4. See also April 1, 1985, pp. CC7–CC8; January 7, 1986, pp. A2–A3; April 22, 1986, pp. CC4–CC5; April 30, 1986, pp. CC1–CC2, G3–G6.
61 *FBIS*, January 8, 1986, pp. CC5, CC8. See also May 29, 1985, p. AA11.
62 Robbin F. Laird, "The Soviet Union and the Western Alliance: Elements of an Anti-Coalition Strategy," IDA working paper, September 1988, p. 10.
63 Ibid., p. 15.
64 In Laird and Hoffmann, eds., *Soviet Foreign Policy*, pp. 364–365.
65 *FBIS*, December 17, 1986, pp. CC2–CC3.
66 *FBIS*, September 9, 1986, pp. CC9–CC10.
67 *FBIS*, September 9, 1986, p. CC10.
68 *FBIS*, October 24, 1986, pp. AA13–AA14.
69 A. Ross Johnson, *The Impact of Eastern Europe on Soviet Policy toward Western Europe* (Santa Monica, CA: RAND Corporation, 1986), p. 64.
70 Gorbachev's interview in *Time* (September 9, 1985), p. 29.

4

Eastern Europe as a Factor in Soviet Foreign Policy toward Western Europe [1]

Charles Gati
Union College and Columbia University

A Conversation in Prague

The author was in Prague on January 28, 1987, the day after Mikhail Gorbachev's major speech "On Reorganization and the Party's Personnel Policy" to the Central Committee of the Communist Party of the Soviet Union (CPSU).[2] The text of Gorbachev's speech, reprinted that morning in the Czechoslovak party daily *Rude Pravo*, seemed to be read with unusual interest by the people of Prague. Its stress on change and the adoption of more pragmatic policies implied change for the better in Czechoslovakia too—in this country of Communist orthodoxy and repression.

Unexpectedly, a leader of the most orthodox faction in the Czechoslovak Communist Party—to be called "Stefan" here—made himself available for a long conversation. A close associate of Central Committee Secretary Vasil Bilak, leader of the hardliners, Stefan is a foreign policy specialist. Known as the Soviet bloc's most vehement critic of Western "imperialism" and especially West German "revanchism," he is also a rigid opponent of all departures from the Soviet model in Eastern Europe.

Without being prompted to do so, Stefan offered a long monologue that morning on how he and Bilak had consistently pursued a constructive policy of good relations with Western Europe, particularly with Czechoslovakia's two neighbors, Austria and the Federal Republic of Germany (FRG). Incredibly, he made it appear that Prague's relations with Vienna and Bonn were all but excellent and that this was due to Bilak's (and his own) efforts to promote intra-European detente between the two halves of Europe. He also suggested that this policy reflected Czechoslovakia's loyal support for the Soviet policy of peaceful coexistence in Europe, notably Gorbachev's wooing of Western Europe.

In fact, because of people like Bilak and Stefan, Czechoslovakia has had the worst relations with Western Europe of all the countries in the Warsaw Pact, marked by frequent incidents along its borders with both Austria and the FRG. Czechoslovakia's trade with Western Europe has remained insignificant, as it had been in the 1970s, because of Prague's extraordinary reluctance to use Western credits, a policy guided by fear of excessive reliance on the West. Put another way, Czechoslovakia did not always follow the ups and downs of Soviet foreign policy toward Western Europe: when relations were poor, Prague followed suit; yet when Moscow sought to improve relations with Western Europe, Prague showed some reluctance to follow suit.

The conversation at party headquarters with Stefan suggested that the Czechoslovak regime in general and the Bilak faction in particular were under some pressure from Moscow to bring their policies in line with the latest Soviet objectives. Stefan's highly defensive monologue was intended to convey the message that such pressure was not necessary, that Czechoslovakia was as always a loyal ally, and that Moscow could count on its support. The kind of *glasnost'* and *perestroika* Gorbachev was recommending to the CPSU might or might not apply to Czechoslovak domestic circumstances, but the foreign policy implications of Gorbachev's statements would be carefully studied and indeed heeded.

Those implications were self-evident to East Europeans. Nuances aside, Soviet foreign policy toward Western Europe entered a highly pragmatic phase in the mid-1980s. Whatever Moscow's long-term objectives, its immediate goals were to lower international tension in Europe and to improve commercial ties between Western Europe and the Soviet Union. The realization of these goals, in turn, called for lowering tensions between Moscow's East European allies, including especially Czechoslovakia and the German Democratic Republic (GDR) on the one hand, and their West European neighbors on the other. Because of Western concerns about human rights, the East Europeans would also have to treat their own people better, reduce repression, and tolerate a measure of diversity. As calculated in Moscow, probably correctly, such East European policies abroad and at home might help convince a skeptical Western Europe of the new Soviet commitment to European detente and thus solicit the kind of response from Western Europe that the Soviet Union is seeking.

Herein lies the main dilemma Eastern Europe presents to Soviet foreign policy toward Western Europe, a dilemma well understood by Stefan and other leaders in the Warsaw Pact. On the one hand, the effectiveness of Moscow's

West European policy depends at least in part on lowering the iron curtain between the two halves of Europe. On the other hand, however, lowering the iron curtain entails risks for Moscow's East European policy, one that aims at protecting the cohesion of the Soviet bloc. In a nutshell, the question is how the Soviet Union can enhance its influence over Western Europe without losing its control over Eastern Europe.

Accordingly, this chapter addresses two questions: (1) In what ways is Eastern Europe an asset, a liability, or both to the Soviet Union in Moscow's attempt to increase its influence in Western Europe? (2) Can the Soviet Union realize Gorbachev's goal of turning Europe into "a continent of peace" without making appropriate political adjustments in Eastern Europe?

The Soviet Conception of Eastern Europe

Ever since the Red Army reached the region at the end of World War II, Eastern Europe has occupied a special place in Soviet foreign policy. Given Joseph Stalin's desire to turn "socialism in one country" into "socialism in one region," Eastern Europe became, and has remained, more than a sphere of Soviet influence. Had it become merely a sphere of influence, Eastern Europe would have been denied to the West only for purposes of close military, political, and perhaps even economic ties. Instead, the East European countries evolved into replicas of Soviet-style socialism. At least at the beginning, Eastern Europe was meant to demonstrate the universal applicability of Soviet patterns and policies.

Among countries at or near the Soviet Union's European borders, Finland, Greece, and Turkey succeeded in retaining their independence. Had the others been allowed to emulate Finland's sphere-of-influence understanding with Moscow, they would have developed carefully balanced foreign policies that would have avoided alienating or irritating the Soviet Union. Because they would have joined no Western military organizations, they would have formed a sphere of cordial and respectful states mindful of Soviet security interests.

Because such an arrangement would have entailed considerable freedom for the East Europeans to fashion domestic political and economic institutions in accordance with local traditions and interests, the region so construed would have been more stable and its people less preoccupied with anti-Sovietism. Under the circumstances, moreover, the Soviet Union would have been seen abroad, not only in Eastern but in Western Europe too, as a great power and a generous neighbor. There can be little doubt that if the Soviet Union had created such a neutral belt of semi-independent states along its Western borders—a sphere of influence, not a sphere of control and domination—its influence in Western Europe would have increased markedly. This would have been the case particularly in Italy and France, countries with strong communist parties whose appeal might not have declined in recent years as much as they have had it not been for Soviet military interventions and other harsh measures applied by Moscow in defense of its domains in Hungary (1956), Czechoslovakia (1968), and Poland (1981).

However, Stalin's original idea of creating not a sphere of influence but a sphere of control has held since the 1940s. From Stalin to Gorbachev, all Soviet leaders have interpreted effective influence to be synonymous with control. They have also come to believe that control over Eastern Europe is synonymous with control over the Soviet Union itself. Following this formula, they could also consider the region's evolution toward socialism as ideologically correct; any significant deviation from this would be but a "historical regression."

Thus, although what the West has called the Brezhnev doctrine was declared only in the aftermath of the Warsaw Pact's 1968 crushing of Czechoslovakia's "Prague Spring," the underlying idea was always, and remains, inherent in the Soviet conception of Eastern Europe. The only difference was that Leonid Brezhnev, unlike his predecessors and successors, offered an explicit definition of how the Soviet Union conceived of its interests in the region and what it would do to protect them. This is what Brezhnev said in November 1968:

> When internal or external forces hostile to socialism attempt to reverse the development of any socialist country and to restore capitalist practices, when the cause of socialism is threatened in that country, the security of the whole socialist community is threatened. It becomes a problem no longer of the people of that country only, but a common problem and concern of all the socialist countries.[3]

The more things (appear to) change, the more they remain the same. Speaking in Prague on April 10, 1987, Gorbachev declared:

> We proceed, first and foremost, from the assumption that the entire system of political relations between socialist countries can and must be unswervingly built on the basis of equality and mutual responsibility. No one has the right to claim a special status in the socialist world. The independence of every party, its responsibility to its people and its right to decide in a sovereign manner how the country should develop are firm principles for us. At the same time, it is our profound conviction that the socialist community cannot succeed unless every party and country looks not only to its own but to common interests, shows respect for its friends and allies and takes account of their interests.[4]

Gorbachev's language seems less threatening than Brezhnev's, but neither in this speech nor in other statements is there any indication of change from past Soviet assumptions, habits, and expectations. In Gorbachev's world too, Eastern Europe can continue to become more diverse since the notion of "different roads to socialism" remains one of the bloc's guiding principles. But the bloc's common interests, identified in ideological terms as "socialist internationalism" and in practice as Moscow's self-assigned right to limit the East European countries' sovereignty, still outweigh other, presumably less desirable principles. In other words, for the Soviet Union Eastern Europe is still a sphere of control rather than merely a sphere of influence.

The foregoing is not meant to suggest the absence of any change in Soviet policies toward the region, only the absence of fundamental change in Moscow's conception of Soviet interests. In fact, recent years—even before Gorbachev—have demonstrated a welcome evolution in Soviet policies, which have come to reflect not only the old hegemonic habits, but also considerable pragmatic flexibility. The new mixture is mainly a product of East European domestic developments, but it has also been dictated by leadership conditions in the Soviet Union and Soviet hopes for improved relations with Western Europe.

First, mainly in response to the region's chronic instability, Moscow has recognized and accepted what might be called "variations on the Soviet theme."[5] On encountering persistent problems, Stalin's successors have been willing to accept adaptations and innovations necessitated by local conditions, traditions, and customs. Their choice was either to rely on coercion, as Stalin had done, and obtain the appearance of bloc cohesion and the reality of instability, or to adopt a more circumspect course, allow for national peculiarities, and thus make the East European regimes more viable—although increasingly different from the Soviet Union and hence more difficult to control. Because there was a price to be paid in either case, Moscow has tended to acquiesce in gradual change so long as the local Communist party's hegemony of power remained beyond dispute.

The second factor moderating Soviet policies has been the deteriorating condition of the East European economies. There was a time when the centrally planned economies in the bloc could at least claim to produce high growth rates. With new investments frequently deferred and the old technologies becoming ever more obsolete, this has not been the case for many years. In general, as shown later in this chapter, Soviet-type economies could not effectively adjust to the requirements of intensive growth, which calls for technological innovation, managerial flexibility, material incentives, and the like.

Recognizing the problem, all East European regimes except that of Romania have tried to experiment with reform, with Hungary and Czechoslovakia's leading the way in the mid-1960s, followed by "streamlining" in East Germany, a few haphazard measures in Poland, and the adoption and minimal implementation of a new economic mechanism in Bulgaria. Only in Czechoslovakia, where economic change eventually led to the 1968 Prague Spring, did Moscow intervene; elsewhere it tolerated economic experiments even when these included, as they did in most cases, an expansion of commercial ties with the West. Clearly, the Soviet Union was more concerned with the region's political stability, which the economic reforms were intended to achieve, than it was with modest deviations from its own economic structures and processes.

Third, a measure of flexibility in Soviet behavior toward Eastern Europe was also a by-product of the Soviet leaders' preoccupation with the struggle for power in the Kremlin. In 1953–55 Nikita Khrushchev extended the processes of de-Stalinization to Eastern Europe, partly to discredit his political opponents, partly to help release regional tension pent up during Stalin's last years. In the mid-1960s, Hungary and Czechoslovakia took advantage of the opportunity for maneuver afforded by a few Soviet leaders' apparent interest in economic reforms and their subsequent absorption in political infighting. In the late 1970s and early 1980s, prior to Gorbachev's ascent, once again it was the struggle for power—the struggle for succession—that preoccupied the Kremlin and permitted Eastern Europe more elbow room.

Fourth, the West in general and Western Europe in particular have been another key factor tempering Soviet policies toward Eastern Europe. Of course, the Soviet Union has not allowed any of its East European dependencies to deny the Communist Party's leading role or to defect from the Warsaw Pact for the sake of Western trade and whatever else detente denotes. When faced with the choice of maintaining its control over Eastern Europe or extending its influence over Western Europe, the decision was invariably in favor of controlling Eastern Europe.

Yet so deep rooted is the East European desire to "belong" to Europe, so magnetic is the West's impact, and so important is Western trade that Moscow has grudgingly allowed for greater intercourse between the two halves of Europe. It has tolerated movement toward intra-European detente in recognition of its own interests in closer ties with Western Europe and of the East Europeans' longing for similar relations. The benefits of West European trade to the Eastern bloc as a whole have also been a consideration. The idea of "one Europe" has remained a pipe dream, but the increasingly rusty iron curtain testifies to the evolution of Soviet policy toward Eastern Europe.

The East European-West European Connection

It is customary to assume that only for economic reasons are the East Europeans interested in the expansion of relations with Western Europe. This is not an inaccurate explanation, but it is incomplete. By comparison with the Soviet Union, Eastern Europe's interests are broader, deeper, and more compelling. After all, Eastern Europe *is* Europe.

The very term *Eastern Europe* is a misnomer. Geographers seldom identified such a region before 1945; when they had, the region so described was the European part of the Soviet Union, the area west of the Ural Mountains. Of the six states comprising what is now called Eastern Europe, four—the GDR, Poland, Czechoslovakia, and Hungary—had been recognized as being in central Europe, while two—Romania and Bulgaria— were in southeastern Europe (or the Balkans).

The map provides ample justification for questioning the accuracy of the current use of Eastern Europe as a geographic entity. For example, while Prague is to the northwest of Vienna, Czechoslovakia is still said to belong to Eastern Europe and Austria to Western Europe. In some accounts nowadays, Greece is placed in Western Europe even though it is located to the southeast of every state that has been assigned to Eastern Europe. Thus, Eastern Europe is no more than a convenient description of the *political* reality of postwar Europe.

Nor are the countries of Eastern Europe culturally very different from their western neighbors. When Europeans think of classical music, they think of Berlin, Vienna, and Prague; when they think of psychoanalysis, they think of Vienna, Berlin, and Budapest; and so on. Despite the similarity between Russian and the other Slavic languages and cultures of Bulgaria, Czechoslovakia, and Poland, the six countries of Eastern Europe do not have much in common with the Soviet Union. The influence of pan-Slavism notwithstanding, Czechoslovak culture and life-styles resemble Germany's more than Russia's, for example. Many aspects of Hungarian culture and life-style remain almost interchangeable with Austria's. In the heart of Europe, political traditions are similar too. Poland's precommunist political institutions could be compared to Finland's, Hungary's to Austria's, and Czechoslovakia's to Germany's (in the 1920s).

To a great extent, the commonality of the European tradition has apparently survived four decades of political division. It is especially strongly felt in the so-called northern tier of Eastern Europe—in what used to be considered the central European countries of the GDR, Poland, Czechoslovakia, and Hungary. So strong is this tradition that some regimes in the region have responded to it not only by lowering the iron curtain for tourists and cultural influences, but also by pursuing policies on intra-European issues that are somewhat different from those of the Soviet Union. In doing so, they have shown keen awareness of their dilemma. In order to rule, they must support Moscow, but in order to govern on the basis of a measure of domestic authority, they must distance themselves from Moscow. Hence, in recent years especially, these nominally loyal allies have tried to inch toward Western Europe without appearing to inch away from the Soviet Union.

Security Issues

Even in the realm of security, where Soviet insistence on bloc cohesion is most persistent, there is evidence of this trend owing to different domestic interests and different reactions to West European sensitivities.

A case in point was the curiously varied response of several Warsaw Pact countries to the deployment of new intermediate-range nuclear missiles (INF) in Western Europe in the mid-1980s. Initially, there was a united and vocal campaign against the planned installation of these INF systems. With the exception of the Romanian position, which blamed both the United States and the Soviet Union for the diplomatic impasse at the Geneva INF talks, the East European campaign closely reflected Soviet themes. East Berlin, Prague, Warsaw, Budapest, and Sofia addressed daily warnings to West European governments about unspecified "dire consequences" and countermeasures if the West proved unwilling to be more "realistic," that is, if it implemented the North Atlantic Treaty Organization's (NATO) so-called dual-track decision. For example, Czech Politburo member and Central Committee Secretary Vasil Bilak warned that his country would fulfill its obligations relating to the defense of the socialist community.

At the same time, however, the East Europeans appealed to the people of Western Europe—"progressive elements" in particular—to think in pan-European terms and defeat this wicked American plan. East Germany's Erich Honecker spoke of the dangers ahead, asking for a "coalition of common sense" in both German states and claiming that he was making his appeal "in the name of the German people." Explicitly relating the missile issue to intra-German relations, Honecker suggested that recent improvements in travel and family reunification matters might have to be sacrificed if West Germany were to accept the new missiles. He also suggested the need for Europeans, especially Germans, to do what they can to limit the damage done to their relations by growing U.S.-Soviet tension.

What was Honecker's motive? Was he assigned by Moscow the task of playing on West German sentiments in order to help defeat NATO's missile decision? Alternatively, was he attempting to tell West Germans of his country's predicament—that the missiles would prompt Moscow to stop the process of ongoing, limited detente between the two Germanies?

It is possible that prior to the collapse of the INF talks in November 1983 there had been collusion or fraternal cooperation between East Berlin and Moscow. That would explain Honecker's careful rhetoric as well as his tolerance of the activities of the unofficial peace movement in the GDR at that time. The purpose of such collusion would have been to convince West German peace activists that their pressure on Washington was being matched by pressure applied on Moscow by East German peace activists. With the situation being portrayed as somewhat similar in both Germanies, West German progressive forces would have gained a measure of respectability among their compatriots.

Granted the possibility of early collusion, it is still not clear why a degree of dissonance between Moscow and East Berlin subsequently continued in both words and deeds. For example, after the INF talks had collapsed, the GDR signaled its interest in European detente by allowing an unusually large number of people to leave for West Germany. In mid-1984, at the height of a virulent Soviet campaign against the FRG for having accepted the missiles, the two German states signed an agreement on new West German bank credits in exchange for East German concessions on emigration, family reunification cases, and the like. If earlier it might have made sense to feign dissonance in the bloc in order to mislead the West, it made sense no longer. After all, the issue of West Germany's accepting the missiles had been settled.

Proof of Soviet irritation with the GDR's *Westpolitik* was provided by a subsequent public debate between the official dailies of the two parties. One ominous article in *Pravda* reminded East Berlin of NATO's "crusade against socialism" and of Bonn's desire to "solicit concessions on matters of principle that affect the GDR's sovereignty."[6] Another *Pravda* piece repeated the same points, but in addition it expressed reservations about various deals made between the two Germanies. This unsigned—and hence fully authoritative—article claimed that West German loans in exchange for East German travel concessions only served to provide Bonn with "new channels for political and ideological influence."[7] Responding to these accusations, the GDR's *Neues Deutschland* asserted that both German states were "independent in their internal and external affairs," a message intended to reach Moscow. The message intended to reach Bonn was that the GDR was still interested in influencing the ongoing debate in the FRG between revanchist and realistic elements because, for the sake of intra-German detente, the East German regime still had a stake in the outcome.[8]

Although similar sentiments were expressed elsewhere in Eastern Europe, only Romania objected publicly to Soviet counterdeployments. In this case too, the perception of common European interests played a key role in Bucharest's decision to defy Moscow. Suggesting that the United States and the Soviet Union were trying to countermand the will of the countries of both

Western and Eastern Europe, President Nicolae Ceausescu argued that the deployment of INF missiles in Western Europe and the retaliatory deployment of new missiles in Eastern Europe were in defiance of public opinion, of the European peoples, and of the whole world.

Thus, the missile controversy generated division within the Warsaw Pact. In the end the East Europeans went along with the Soviet decision to counter NATO's new missiles with the deployment of new Soviet missiles, but they did so grudgingly. The price of military gain for the Soviet Union was a measure of political dissonance within the region. The East Europeans, ultimate willingness to accept the Soviet missiles was due to Soviet pressure and perhaps a sense of internationalist obligation. Their reluctance to support Moscow more enthusiastically was primarily due to their opposition to policies that would damage chances for improving relations with Western Europe and therefore for intra-European detente.

Political Issues

In the realm of politics and ideology too, it is mainly in times of tension between the superpowers that the East Europeans voice most strongly their concern for pan-European cooperation. The measures they undertake on such occasions are largely symbolic, but they suggest an undercurrent of dissatisfaction with Moscow's approach to Western Europe. In the aftermath of the missile controversy, for example, Hungary's Janos Kadar held summit meetings with several West European heads of state in Budapest and visited France—at a time when Moscow refused to receive high-level West European delegations.

Indicative of some East European political attitudes was, and remains, the redefinition of the role of small- and medium-sized East and West European states in the East-West dialogue. The issue first surfaced in January 1984 when U.S.-Soviet relations were characterized by references to "Cold War II." In the Hungarian party's ideological monthly *Tarsadalmi Szemle (Social Review)*, Matyas Szuros, the Central Committee secretary in charge of foreign policy, published a blunt article calling for an end to the "violation of the principle of equal rights and independence" within the bloc.[9] Szuros argued that in the absence of a final arbiter in world communism to develop a common strategy for all, each Communist party must strike a balance between policies that serve common interests and policies that reflect individual or national interests. He implied that, when faced with an especially difficult choice, a Communist party might occasionally pursue national interests at the expense of internationalist duties.

The point of Szuros's dissertation and the meaning of these code words became clear when he volunteered that the small- and medium-sized European states, East and West, should help the two superpowers work out "mutually acceptable and rational compromises." The meaning was unmistakable—and heretical. The new U.S.-Soviet cold war need not signify East-West cold war in general and a confrontation in Europe in particular. Moreover, what made the Hungarian position important was that it was supported by the GDR and especially by Romania. A leading member of the Romanian Communist Party's Executive Committee (Politburo), Dumitru Popescu, declared that the criterion of ideology loses all significance when it comes to the issue of war and that therefore Romania does not believe itself obligated automatically to identify itself with a bloc policy of tension and confrontation.

Without mentioning Hungary, Romania, or East Germany, the Czechoslovak party issued a stern and immediate rebuttal. In an article published in the official daily *Rude Pravo*, two high-ranking Czechoslovak loyalists defended the need for a common front against Western imperialism. Sarcastically, they weighed in against those who "have been inclined to 'demonstrate' a kind of 'independence' in foreign policy, deviating in this way from the line that was agreed by the [socialist] community as a whole" and those who "have been musing about 'the role of small states' who should allegedly assist the 'great powers' in coming to an agreement."[10] More than a year later, *Pravda* joined the debate with an especially militant article. "What question can there be," the pseudonymous author asked rhetorically, "of any mediation by particular socialist countries in resolving disagreements between the USSR and the United States if on key international questions the foreign policy of the USSR and [Eastern Europe] is identical?"[11]

Even those unfamiliar with communist jargon could easily understand the issue in dispute. Those East European states that had developed a vested interest in intra-European cooperation—notably Hungary, Romania, and the GDR—were attempting to protect European detente from the vicissitudes of U.S.-Soviet relations. They justified their position by distinguishing between international and national interests and by assigning an important role to small- and medium-sized states in both Eastern and Western Europe. This position, in their opponents' view, was heretical because, according to traditional Marxist analysis, the nature of class rule rather than a state's size should reveal its essential features and interests. Thus, those proposing a special role for small- and medium-sized countries suffered from a lack of appreciation for differences between the socialist states of Eastern Europe and the "bourgeois" states of Western Europe.

Economic Issues

Of course, economic considerations have also pulled Eastern Europe toward Western Europe and away from Moscow. Even the political-ideological dispute just decribed can be explained in part by economic considerations, for the moderate position toward the West was adopted precisely by those countries whose economies have had the greatest Western exposure—Hungary, Romania, and the GDR. Conversely, Czechoslovakia—alone in the Soviet bloc—has tried to protect itself from what it regards as dangerous reliance on the West; its Western debt, for example, has been almost completely nonexistent.

For Eastern Europe as a whole, the Soviet economic connection is a vitally necessary, if insufficient, source of further growth, productivity, and effectiveness. Its value is twofold: First, because the region is energy poor and lacks hard currency, only the Soviet Union can fulfill Eastern Europe's vast energy needs. Second, the Soviet Union offers a huge market for such East European products as industrial machinery, equipment, food, and consumer goods for which a ready market might not exist elsewhere in the world.

What made the Soviet Union not only a necessary but also an advantageous trading partner in the 1970s was Moscow's willingness to subsidize its energy sales to the region. Although Western analysts disagree about the size of Soviet subsidies, they do agree that the subsidies reached billions of dollars or rubles per year during that decade.[12] Fearful of political instability, Moscow offered such subsidies in order to reduce the immediate impact of the sudden and substantial increase in the world market price of energy, particularly oil.

With the sharp decline in subsidies since 1981, the Soviet Union has become a less desirable, if still essential, trading partner for Eastern Europe. Although the Soviet decision to reduce and even eliminate the subsidies was understandable in view of the contraction of the Soviet economy itself, the East European regimes were, and remain, disappointed. They were counting on these subsidies to accelerate economic growth and satisfy popular demands for a higher standard of living. Their disappointment was equally understandable, for they had made an implicit contract with Moscow according to which they would remain supporters of Soviet policies and keep their part of the bloc stable, in exchange for which Moscow would protect them from the worst effects of international economic turbulence.

The resulting slowdown of the East European economies in the 1980s left them with no realistic alternatives. The East Europeans could not afford to reorient their trade toward Western Europe at the time of greatest need to do so and when several states in the region were politically prepared to do so. Because they had wasted so much of what they borrowed in the 1970s and thus all but eliminated the possibilities of obtaining additional Western credits,[13] they could not acquire Western technology to make the transition from extensive growth (that is, quantitative increases achieved largely by labor and capital investments) to intensive growth (that is, qualitative improvement achieved largely by the infusion of high technology and more efficient management techniques). Nor could they buy Western consumer goods the region's people were so eager to have.

Although East European statistics are often unreliable, the available figures clearly indicate the trend. For example, while imports from the West increased more than fivefold between 1970 and 1980 (from $5.25 billion in 1970 to $27.45 in 1980), they actually declined more than 30 percent between 1980 and 1985.[14] Exports to the West also increased about fivefold between 1970 and 1980, but since 1980 they have remained basically unchanged at about $25 billion. True, these figures are strongly influenced by Romania's extraordinary austerity measures, which include the policy of almost no Western imports until that country's debt is paid off, and by Poland's inability to pay for badly needed Western imports. Still, there is a general problem in Eastern Europe as well. The region's economies cannot penetrate Western markets unless new technologies are obtained and properly utilized, but these new technologies cannot be obtained unless they are paid for from the sale of East European products to the West. It is, indeed, a "Catch-22" situation.

With West European economies on the upswing in the 1980s, the East European economies have experienced a long period of stagnation in this decade. Growth rates hovered around a meager 1 percent per year in the first half of the decade (compared to a healthy 5 percent in the first half of the 1970s), and the prospects are gloomy. For obvious reasons, the FRG is still deeply interested in being helpful to East Germany. Several West European governments and banks, as well as Japan, are still prepared to extend credit to reformist Hungary. Nevertheless, for economic and political reasons, the further extension of intra-European trade is unlikely. The main economic stumbling block is the size of Eastern Europe's hard currency debt. The political stumbling block is the fear shared by both the Soviet Union and some of the East European regimes of radical economic reforms—the prerequisite of viability and stability—that would sooner or later entail political reforms as well. Yet in the absence of such reforms Western Europe will be reluctant to expand its economic relations with its Eastern neighbors.

Continuity under Gorbachev?

All in all, the long-term trend in intra-European relations is still positive. After all, who would have thought 15 or 20 years ago that West Germany would become the second largest trading partner of every East European country? Who would have thought that Hungary would eventually abolish the visa requirement for citizens of neighboring Austria? Who would have thought that the GDR would build a cable television system so that the people of the city of Dresden, which is surrounded by hills, could watch West German television as their compatriots can and do everywhere else throughout East Germany?

Over the years intra-European relations have become more complex and the East European conception of Western Europe has become more realistic. Today, the region's Communist regimes assume that capitalist Western Europe, allied with the United States, is here to stay. There is almost no fear of war originating in the West. Responding to popular longing for "belonging" to Europe, most of official Eastern Europe has in fact moved markedly ahead of the Soviet Union in trying to lower the iron curtain between the two halves of Europe.

As far as Moscow's West European policy is concerned, Eastern Europe is still a problem and indeed a major liability. True, Eastern Europe makes possible the forward deployment of Soviet military forces and weapons, a military advantage in case of war and a political advantage insofar as the Soviet Union seeks to intimidate Western Europe. There is something to be said for East European contributions to the Warsaw Pact, including their modernization efforts in recent years. There are also advantages to be derived for Moscow from the relative ease with which East German intelligence can penetrate the FRG and other German-speaking areas of Europe, obtaining and sharing with Moscow valuable political information and technological data from the West. Yet, on balance, the difficulties Moscow has repeatedly encountered in Eastern Europe, and especially its interventions, represent a significant handicap for the Soviet Union's West European policy.

Fear of losing control over Eastern Europe is the main reason Soviet policy has long sought to isolate and intimidate rather than to seduce Western Europe, especially the FRG. Simply put, the alternative policy of making itself appealing to (and thus trying to seduce) Western Europe would have entailed a process of "historical regression" in Eastern Europe. Had the Soviet Union opted for that alternative, it would have had to proceed first by transforming Eastern Europe from a sphere of domination to a sphere of influence. It would have had to allow East European membership in Soviet military and economic alliances to be coupled with internal and even external independence for these nations—much the way another East European country, Greece, has come to relate to the West.

Unwilling to accept a Greek formula for Eastern Europe, Soviet policy still aims primarily at isolating Western Europe from the United States. That approach has been implicit in Gorbachev's various arms control proposals, which—important as they are—represent a substitute for rather than a step toward the resolution of the continent's political division.

Whether Gorbachev will transcend the traditional Soviet approach by attempting to overcome the artificial division of Europe by political means remains to be seen. Several of his East European allies, such as Hungary and probably Poland, would cautiously encourage Gorbachev to undertake a comprehensive political approach to Europe, one that would presumably include the adoption of the Greek formula for Eastern Europe. Yet since Soviet policy toward the two halves of Europe has long proceeded from the motto of "what's mine is mine, what's yours is negotiable," it is as yet premature to expect the Soviet Union to exchange control over its domain for potential influence in Western Europe.

Notes

1 Research for this chapter was completed in May 1987. For a more detailed analysis of Gorbachev's dilemmas in Eastern Europe, see the author's "Gorbachev and Eastern Europe," *Foreign Affairs*, Vol. 65, No. 5, Summer 1987, pp. 958–975. See also Gati, "Eastern Europe on its Own," *Foreign Affairs: America and the World 1988–89*, Vol. 67, No. 3, pp. 99–119.
2 "On Reorganization and the Party's Personnel Policy; Report by Mikhail Gorbachev, General Secretary of the CPSU Central Committee on January 27, 1987," *Moscow News*, Supplement to Issue No. 6 (1987), pp.1–12
3 *Pravda*, November 13, 1968.
4 *Pravda*, April 11, 1987.
5 For further details on this point, see the author's "The Soviet Stake in Eastern Europe," in Seweryn Bialer and Thane Gustafson, eds., *Russia at the Crossroads: The 26th Congress of the CPSU* (London: Allen & Unwin, 1982), pp. 178–191.
6 Lev Bezymenskii, "In the Shadow of American Missiles," *Pravda*, July 27, 1983.
7 "On the Wrong Track," *Pravda*, August 2, 1984.
8 *Neues Deutschland*, August 1, 1984.
9 Matyas Szuros, "The Reciprocal Effect of the National and the International in the Development of Socialism in Hungary," *Tarsadalmi Szemle*, January 1984, pp. 13–21.
10 Michal Stefanak and Ivan Hlivka, "The National and the International in the Policy of the KSC," *Rude Pravo*, March 30, 1984.
11 Vladimirov, "The Main Factor in the World Revolutionary Process," *Pravda*, June 21, 1985, pp. 3–4. "Vladimirov" was widely believed to have been the pseudonym for Oleg Rakhmanin, then first deputy chief of the Socialist Countries Department of the CPSU's Central Committee. Rakhmanin has since lost all of his positions.
12 Compare Michael Marrese and Jan Vanous, *Implicit Subsidies and Non-Market Benefits in Soviet Trade with Eastern Europe* (Berkeley: University of California Press, 1982), and Paul Marer, "The Political Economy of Soviet Relations with Eastern Europe," in Sarah M. Terry, ed., *Soviet Policy in Eastern Europe* (New Haven and London: Yale University Press, 1984), pp. 155–188.
13 For additional data and analysis, see Ivor C. Coffin, "East European Debt Problems and Prospects for Trade with the Developed West," in Philip Joseph, ed., *The Economies of Eastern Europe and Their Foreign Economic Relations* (Brussels: NATO, 1986), pp. 311–329.
14 Compare Roger E. Kanet, "East European Trade in the 1980s: Reorientation in International Economic Relations", in ibid., pp. 291–310.

5
Soviet Perspectives on French Security Policy[1]

Robbin F. Laird

This chapter examines Soviet analysts' perceptions of French security policy since the establishment of the independent nuclear force in the mid-1960s. Soviet analysts argue that General Charles de Gaulle's original policy of independence has been progressively modified. The French are perceived to be trying to bolster the West European component of the Western Alliance in order to strengthen the Alliance.

Soviet perceptions of the French force structure are also analyzed to determine Soviet views on the independence of French forces. Soviet analysts tend to argue that French nuclear forces are, by and large, operationally independent of other Western forces. Politically, however, they are considered to be an integral component of the Western Alliance's nuclear potential. In contrast, French conventional forces are considered to be integral elements of NATO's military potential in Europe, both operationally and politically, regardless of French pretensions of independence. Prudent Soviet military planners would clearly plan to operate against French conventional forces in time of war. There is, however, some disagreement in the Soviet literature over when France would use its nuclear forces. Some think France would use nuclear forces solely to defend French territory; others believe these forces would be used before French territory was invaded. Prudent Soviet military planners would undoubtedly plan, at a minimum, to conduct conventional military operations against French nuclear forces, on land as well as at sea.

Finally, Soviet perceptions of the political and military impact of French security policy are assessed. It is clear that Soviet analysts are more concerned with the political impact of French security policy (including the French nuclear forces) than with its military repercussions per se. The major political concern for the Soviets is the encouragement France has given to the emergence and growth of an independent West European power center. Although France's challenge to American hegemony in Western Europe is significant to Soviet calculations about American power, it is France's contribution to the West European power center that worries the Soviets more.

The Evolution of French Security Policy

Soviet analysts perceive that French security policy has gone through three phases of development since France created its independent nuclear force. The first phase (the 1960s) established the French policy of independence associated with Charles de Gaulle. The second phase (the 1970s) adjusted the original Gaullist system, especially by President Valéry Giscard d'Estaing in the mid-1970s, to accommodate West European and American interests better. The third phase (the 1980s) has accelerated the processes of the 1970s. President François Mitterrand has more fully embraced West European security interests and has promoted a partial return to Atlanticism.

President de Gaulle (1958–1969)

General Charles de Gaulle sought to create a security policy for France that would give greater independence than had been provided by the Fourth Republic's close ties with the Americans in the NATO alliance. De Gaulle tried unsuccessfully to reorganize the Western Alliance to give France a greater role in Western decision-making. As a result of this failure, de Gaulle made the decision to withdraw French forces from the integrated military command dominated by the United States. Soviet analysts hoped that France's withdrawal from NATO's integrated command would be the beginning of the end for the Alliance.[2] Although such hopes were short lived, the political and symbolic impact of the French action has not been. For Soviet analysts, the sine qua non of French independence was, and remains, the withdrawal from NATO's integrated military command.[3]

The symbolism of President de Gaulle's anti-American actions has had a deep influence on Soviet analysts, inducing them to look largely with favor on him and his historical legacy. Moreover, de Gaulle not only distanced himself from the United States but also began to develop a "special relationship" with the Soviet Union. De Gaulle is seen by Soviet analysts to have promoted Soviet-French cooperation in economic, diplomatic, and political affairs; his motivation was to enhance the prestige of France as a world power.[4]

De Gaulle's belief that NATO could no longer effectively guarantee France's security contributed to the country's need for an independent nuclear force.[5] The French nuclear force was declared to be designed to deal with threats from all directions, although the Soviets knew full well against whom it was directed. Nonetheless, the main purpose of the French nuclear force was political, not military; it was seen to be needed to create a political space for independence. As noted in a 1963 article, "In de Gaulle's plans the atomic bomb is not just a weapon in the arsenal for war against the Soviet Union, but also *an instrument of blackmail with regard to the 'allies' of French imperialism.*"[6]

Nonetheless, Soviet analysts did note that France maintained its treaty commitments to NATO and continued to rely on American involvement in West European defense. The result was the creation of an additional threat to the Soviet Union as de Gaulle aspired to build a powerful Western Europe. The Soviets, like de Gaulle, have been skeptical of Europe's ability to develop transnationally through an organization like the European Economic Community (EEC). De Gaulle's desired path to strengthen Western Europe was to establish a directorate of the major West European powers. Although at the time of de Gaulle's presidency French actions increased tensions among West Europeans, Soviet analysts continued to be concerned with the long-term potential for France to stimulate West European cooperation.[7]

President Pompidou (1969–1974)

De Gaulle's immediate successor, Georges Pompidou, was perceived to have gained ascendency primarily as the intellectual heir of de Gaulle. Under Pompidou, France continued to distance itself from American foreign policy. According to Soviet analysts, Pompidou promoted Europeanism as opposed to Atlanticism within Western Europe, namely through "independent European cooperation."[8]

As part of a quest for a separate political space, France under Pompidou deepened its special relationship with the Soviet Union, including economic, scientific, and technical cooperation.[9] Under Pompidou, France also began to strengthen its relationship with the Arab world, thereby providing another area of common interest between France and the USSR. French arms sales and economic ties to Arab nations increased, leading to conflict with the United States, which maintained its close ties to Israel.

President Giscard d'Estaing (1974–1981)

Valéry Giscard d'Estaing continued to deepen cooperation with the other European states, creating a second phase of French security policy in which France moved closer to the concerns of its European neighbors. Giscard, unlike de Gaulle, worked effectively with the EEC rather than condemning it as a supranational force undercutting French sovereignty. At the core of French cooperation with the EEC was the development of the Paris-Bonn axis to the West European power center. The President of France and the Chancellor of West Germany were especially active in the 1970s in promoting economic and political cooperation between the two countries. According to V. P. Slavenov in his comprehensive analysis of the presidency of Giscard d'Estaing, "The politico-economic Paris-Bonn tandem, while remaining a driving force in 'European development,' has also become a main factor in the Community's opposition and resistance to attempts at an American diktat in the world and, especially, in economic affairs."[10]

Despite an emphasis on the European dimension in his foreign policy, President Giscard d'Estaing also reduced the level of tension with the Americans. By the mid-1970s the United States and France were avoiding sharp confrontations with each other, for both sides needed to work together more closely to deal with the deepening "crisis of world capitalism". One critical sign of this new posture was the altered French stance toward U.S.-European security relations. Although continuing to insist on the primacy of its nuclear weapons as a force for independence, France embraced a closer relationship with the Western Alliance on defense issues. Soviet analysts have concluded that by the mid-1970s France's military posture had clearly changed. According to T. I. Sulitskaya, "beginning in 1975 Paris implemented a number of serious compromise measures that were aimed at stabilizing its ties with the NATO military organization."[11] Moreover, A. Kovalev noted statements by General Guy Méry, the French Chief of Staff, that implied a change in French doctrine. "For the first time in the 18 years of the Fifth Republic's existence it was declared that the French armed forces would participate in a 'battle' ... according to the NATO concept of the 'advanced line of defense,' which France had refused to adopt until just recently."[12] Especially indicative of France's changing position in the East-West military competition was the promulgation of the Military Program Law of 1977–1982, which included plans for significant nuclear modernization.

In spite of the partial rapprochement in French-American relations, tensions remained. France wished to use its improvement of relations with

NATO to strengthen the West European power center. The Americans simply wished to reassert Atlanticism in a new form, namely, "interdependence" or "trilateralism."[13] France's efforts to strengthen West European independence were seen to lead inevitably to confrontation with the United States.[14] The United States could no longer rely on Atlanticism to dominate Western Europe. The West Europeans now realized that they "were an important component of the 'security' of the Americans," according to Yu. Davydov.[15] This realization provided the West Europeans with more room for maneuver with the Americans.

For Soviet analysts, a critical dimension of French independence continued to be France's policy of detente toward the Soviet Union. Annual Franco-Soviet summit meetings were held throughout the presidency of Giscard d'Estaing. One Soviet analyst went so far as to suggest that French influence within the West European power center was due largely to its "realism" in dealing with the Soviet Union.[16] As U.S.-Soviet relations worsened in the late 1970s, France was seen as an important bulwark for detente in Europe. When Yu. Davydov argued that Western Europe and the United States had developed different interests in detente, he had in mind the role of France in maintaining the European interest in detente. By the "early 1980s France and the Soviet Union had solid experience in fruitful cooperation in the interest of peace."[17] Nevertheless, all was not rosy in Franco-Soviet relations. In fact, the changes in French policy introduced by Giscard d'Estaing—designed to increase France's role in European security—were seen by the Soviets as basically incompatible with detente.

President Mitterrand (1981–)

The defeat of President Giscard d'Estaing by François Mitterrand in the 1981 election was due, in Soviet eyes, to the economic difficulties confronting France.[18] It was expected that the new government—an alliance until July 1984 of the French Socialists and Communists—would continue the "positive" elements of Giscard d'Estaing's policy, namely, the special relationship with the Soviet Union. Detente would be bolstered in the difficult times of the early 1980s by continuation of the special relationship. Instead, France toughened its stance toward the Soviet Union and departed from the detente policy of the past.

A. Kudryavtsev illustrates the Soviet view of the deterioration in French foreign policy. Although French officials continue "to repudiate decisively the possibility of France's returning to the NATO military organization, at the same time they emphasize allegiance to allied duties."[19] The Mitterrand administration has interpreted as part of its allied duties the propagation of the line that the Soviet Union has gained a military edge in Europe. Mitterrand has called for "more arming to strengthen the U.S. nuclear guarantee" by supporting the deployment of U.S. Pershing IIs and cruise missiles in Europe. But "it is hard to understand how erecting a nuclear fence very near France's eastern borders will strengthen its security."[20]

Despite their hopes, the Soviets recognize that concrete Franco-Soviet relations deteriorated during the 1980s. The French government not only began to downplay the Franco-Soviet economic relationship, it also clearly identified the Soviet Union as the military threat against which French forces are deployed for the first time in the Military Program Law adopted in late 1983.[21] Soviet analysts, however, still hoped that France would reverse its drift into anti-Sovietism.

The drift toward anti-Sovietism was accompanied by a revival of Atlanticism. The main elements of this revival were French support for the deployment of U.S. intermediate-range nuclear (INF) missiles in Europe, greater cooperation with the Americans in military affairs, and a noticeable shift in the French doctrine of independence toward greater cooperation with NATO.[22]

The Soviets considered French support for INF deployments to be threatening to Soviet interests, even though none of these missiles were deployed on French soil. The Mitterrand administration's position on this issue also contradicted the traditional French position of independence.[23] In addition, the French government, by refusing to allow its strategic forces to be included in the INF negotiations, "objectively" supported U.S. interests. Soviet analysts consistently underscored that French identification of the Soviet Union as the threat against which French missiles are directed, as well as France's continued adherence to the Western Alliance, means that French nuclear forces are part of the NATO potential.[24] The Mitterrand administration is also perceived to have stepped up the level of direct military cooperation with the Americans, for example by allowing U.S. nuclear submarines to call at French ports and permitting planes from U.S. aircraft carriers to fly over French territory and refuel in French air space.

Soviet analysts have particularly noted the deepening cooperation between Bonn and Paris in security affairs. In the opinion of Soviet analysts, Paris has drawn closer to NATO in order to work more closely with Bonn.[25] The two countries have deepened their collaboration in military production (such as the joint combat helicopter) and in military-political affairs (such as the joint Franco-German brigade and regular staff talks). In fact, the strengthening of the Bonn-Paris axis in security affairs has been seen by Paris (and Moscow) as a means of building greater European defense efforts within the Western European Union (WEU). This also relates to the Soviet concern that France's relationship with West Germany might encourage negative tendencies in West Germany. Specifically, France encouraged the WEU to lift the remaining restrictions on the joint production of offensive arms with West Germany.[26]

Associated with the shift toward Atlanticism has been a shift in the French force posture. The basic alteration most noted by Soviet analysts is the formation of the Force d'Action Rapide (FAR, or Rapid Action Force). This 47,000-person mobile force is capable of being deployed from French soil to overseas areas or to the forward positions of NATO in the event of a European war. For the Soviets, the deployment of the FAR is inextricably intertwined with NATO's operations, at least in the European theater. As Yu. Buskin noted:

> The possibility of "concerted action" between the French units and the NATO forces in the center of the European continent... has already become a reality. General de Llamby, to whom the mission of building the FAR has been entrusted, told journalists that from now on the question is one of elaborating jointly with the allies the "procedures for actions" for these forces in the European theater. In particular, the FAR will have access to the fuel and ammunition depots, including anti-tank missiles, on the territory of the NATO member countries. The question of ensuring the immediate transfer of FAR infantry units across allied territory has also been examined. Is it not greater than simply technical and tactical cooperation between French Army units and the armed forces of their NATO allies?[27]

In short, the departures from de Gaulle's policy of strict independence made by Giscard d'Estaing have been accelerated under the Mitterrand administration.[28] One Soviet analyst characterized France as having a "new defense strategy" under Mitterrand. There has been "a switch from de Gaulle's 'omnidirectional defense' to 'defense in forward positions.' French statesmen have talked much about France's indisputable membership in NATO and the consequent readiness to fulfill its allied obligations."[29]

Nonetheless, serious tensions remain in French relations with NATO. France continues to exercise full sovereignty over its nuclear weapons. France under Mitterrand emphasizes the primacy of nuclear weapons and, unlike the Giscard administration, has reduced expenditures for conventional arms. This heavy emphasis on the use of nuclear forces has tended to weaken France's ties to the NATO bloc, one Soviet analyst observed.[30] Also, France continues to work against American hegemony in the Western Alliance and toward strengthening West European independence. This was evident, for example, when France rejected the American position in the dispute over the West European pipeline contract with the Soviet Union.[31]

As has been the case in Soviet relations with many countries, the entry of Mikhail Gorbachev on the scene has now infused new life into Franco-Soviet relations.[32] The Soviets no longer consider France to be following an exclusively anti-Soviet course of strengthened Western Alliance relations. Starting with the renewel of regular meetings between French and Soviet leaders in the summer of 1984, and especially with Gorbachev's visits to France in 1985 and 1989, Soviet analysts view relations with France as having improved, albeit inconsistently. In the first two years of Gorbachev's tenure, these bilateral relations were characterized as positive, realistic, and balanced, terms that would not have been applied before 1984.[33] For instance, in contrast to Mitterrand's 1984 meeting with Soviet officials in Moscow, when he made one of the most forceful presentations of Western interests on the human rights issue in memory, the speech Mitterrand made during his 1986 trip incorporated a much milder statement on this topic. This visit also indicated to the Soviets that the conservative victory in Parliament—resulting in the formation of a cohabitation government with President Mitterrand and Prime Minister Jacques Chirac—had not profoundly altered France's improving relations with the Soviet Union. Thus, Soviet commentators initially perceived no significant differences between Chirac and Mitterrand on foreign policy matters.

Yet even this reprieve, when Soviet analysts viewed France to be following a more independent (and less pro-United States) policy in East-West relations, proved temporary. In late 1986, and especially in 1987, bilateral Franco-Soviet relations again experienced a notable chill. In part, France's cohabitation government was seen to adopt a more anti-Soviet and pro-Atlanticist position. The French stance on nuclear weapons (as well as on human rights) caused the greatest amount of tension between the two countries. As U.S.-Soviet progress at the arms control negotiating table continued, French opposition to this progress intensified. Even the Soviet decision not to insist on the inclusion of French and British nuclear weapons in the INF talks did not assuage French concerns about these negotiations. Accusations of France's short sightedness in its approach to nuclear arms control became commonplace in the Soviet press. Obviously these disputes could not help but influence the evolution of bilateral relations as well. The differences in Gorbachev's assessment of the situation during Mitterrand's visit to Moscow in July 1986 and Chirac's visit in May 1987 highlight his greater caution on the latter occasion. To Mitterrand, Gorbachev commented that "the Soviet Union has firm intentions to widen the areas of agreement and cooperation between the USSR and France, to do all that is necessary so that the Soviet-French dialogue will once again become the generator of healthy tendencies in world politics."[34] In contrast, during Chirac's stay, Gorbachev stated that new "impulses" were needed in order to return to the countries' previous friendly state of relations. Although it was hoped that the Chirac visit would ease some of the friction between the two countries, at the end of the visit the Soviets continued to criticize French policy, especially on disarmament. Yegor Ligachev carried the theme of needing a new impulse in relations to France on his trip in December 1987, thereby indicating that not much progress had yet been made.

The 1988 presidential election in France was heralded as having important international implications. Specifically, V. Bol'shakov argued that Mitterrand's defeat of Chirac was connected with the French people's rejection of Chirac's policy of a super-arms race, which Mitterrand's socialist party had tolerated during cohabitation. Bol'shakov clearly differentiated between Mitterrand and Chirac on foreign policy. In particular, he explained that changes in Soviet policies have forced Mitterrand and the socialist leadership to reexamine their previous defense priorities. Whereas Chirac virtually ignored completely U.S.-Soviet progress in arms control, Mitterrand announced soon after his reelection that he would do everything necessary for the cause of peace and disarmament[35] and subsequently voiced his support for Gorbachev's peace initiatives. Mitterrand further announced that he would meet with Gorbachev by the end of the year in order to reestablish regular meetings between the two leaders.

A meeting did, in fact, take place in November 1988 in Moscow. *Pravda* expressed the hope that bilateral cooperation would be improved and that the two nations' positions on world issues—including disarmament and East-West relations—could be brought closer together. During the visit, both leaders placed considerable emphasis on the importance of continuing the political dialogue, recognizing that the full potential of Franco-Soviet relations has not been realized, either politically or economically. Yet while they did refer to problems in relations over the past few years, more attention was turned to hopes for the future. In this connection, the joint Franco-Soviet space flight was seen to reflect such hopes; the leaders praised the undertaking as a possible signal for a new stage in the development of bilateral cooperation.

Although Franco-Soviet bilateral relations clearly have seen better times than during the 1980s, there have at least been some areas of common interest. First, the resumption of relatively regular high-level visits between Mitterrand and Gorbachev has, at a minimum, provided the opportunity for personal contact between the leaders.[36] Of particular significance was the fact that Gorbachev's first visit to a Western country after assuming control was to France. Mitterrand's most recent visit to Moscow would seem to indicate that the worst period in Franco-Soviet relations has now passed.

Another area in which Soviet commentators have found some common ground with France concerns military-political affairs. Primarily at issue here is the Strategic Defense Initiative (SDI). France's open refusal to participate in a project so dear to the U.S. administration is continually noted by the Soviets. In a comprehensive article, A. Kudryavtsev observed that France had "objective interests" in preventing the militarization of space owing partly to the fact that the system would likely protect only the United States, not Europe. Kudryavtsev further noted:

> If because of Washington the arms race spreads into the area of antimissile systems, then [France's] national military system would be faced with a dilemma difficult to resolve. Either the deterrent capability of French nuclear forces will be radically lowered, and the thirty year effort to create them would seem, in effect, unneccessary. Or it will be necessary to enter into the next round of growing military expenditures—an option quite undesirable, considering the high cost of producing modern weapons and the long-term decrease in economic growth rates. The "Strategic Defense Initiative," if it were realized, could serve as a detonator of an arms race of such proportions that, prospectively, as it is feared here, it could put into question the financial and even technical capabilities of the country to retain its place and role as a great power.[37]

Kudryavtsev also cited the French fears that it might lose its leading role in Europe to West Germany through the latter's participating in SDI.[38] In order to avoid falling behind the technology generated by SDI, France has initiated "technological diplomacy" through the EEC's space program and the European-wide scientific-technological research program known as Eureka.[39]

The ups and downs of Franco-Soviet relations are, of course, further compounded by France's continued membership in the Western Alliance. The French freely admit their membership in the Western bloc of nations, but within this broad framework, they have steadfastly continued to proclaim their independence; nowhere was this stance better illustrated than in the French refusal to include their nuclear forces in the U.S.-Soviet INF negotiations.

In short, France remains an independent player in the East-West competition. France is shifting from the more dramatically independent position taken in the days of de Gaulle as it seeks to play a major role within the Western Alliance. Rather than rejecting the Western Alliance, the French are now perceived by the Soviets to be working against American interests in dominating the Alliance. The tensions between the French effort to strengthen the European pole of the Alliance and the U.S. effort to use its Western allies to implement American national security objectives will be of increasing significance in the years ahead. Yu. Davydov has expressed this tension: "The U.S. has set for itself ambitious objectives for which it does not have the resources, at the same time as Western Europe has determined that its foreign policy objectives must proceed from real, not imaginary, national (or regional) potential."[40]

The French Force Posture

The Soviet view of the evolution of French security policy emphasizes both the relative independence of French policy and its significance for the

Western Alliance. The latter element is addressed in the next section. Here, the focus is on Soviet perceptions of the French force posture and its meaning in relation to French security policy. Put in other terms, how independent are the French forces perceived to be, politically and militarily?

The core of France's policy of independence is its nuclear force, a fact consistently noted by Soviet writers. In rankings of missions for the various French forces, the nuclear mission is always ranked highest by Soviet analysts, especially for the French navy and air force.[41] Only the army is given a more ambiguous mix of missions, with the nuclear mission much less clearly in command of resources. The army is, however, trained to use, and to exploit the use of, tactical nuclear weapons.[42]

Soviet analysts perceive French nuclear forces, although an expression of a desire for independence, to be integral components of NATO's nuclear forces because of France's continued commitment to its Alliance treaties. As I. A. Koloskov noted in 1976, "Although serious contradictions have arisen between France and the United States and between France and NATO, the leaders of the Fifth Republic have never—even during the years of deteriorating Franco-American relations—cast doubt on their alliance with the United States and their participation in the Atlantic Pact."[43] France's solidarity with the Alliance was underscored frequently in Soviet commentary on the Euromissile crisis as a reason for inclusion of French forces in the INF talks.

Militarily, the situation is more ambiguous. French forces have an independent cast in some respects. Most significantly, France exercises full sovereignty over the decision to use its nuclear weapons; that is, France can make an independent decision to use nuclear weapons.[44] France, unlike the United Kingdom, has developed its nuclear forces largely independently of the United States. Nonetheless, Soviet analysts argue that to execute its nuclear strike, France relies on NATO to some extent, particularly in the close connection between its surveillance and warning systems and NATO's corresponding system.[45] French targeting consists not only of Soviet cities, but also of military and administrative-industrial targets crucial to any Soviet war effort.[46]

Soviet analysts consider French nuclear forces and the doctrine governing their use to have become less independent as French security policy has evolved; they are now less certain that there is a strict correlation between nuclear deterrence and the defense of French territory. In an authoritative treatment of French military doctrine, Captain A. Karemov and Colonel G. Semin argued that France had adopted its own version of the flexible use of nuclear weapons. Whereas before the mid-1970s French officials spoke of the use of strategic nuclear weapons in a massive response, since then they have talked of using nuclear weapons "in stages" as France's vital interests are threatened.[47] The significant enhancement of French nuclear forces in the years ahead is seen to be in support not of a massive retaliation doctrine, but of a doctrine of nuclear weapons used in stages. As an authoritative article in *Pravda* underscored:

> The present French military doctrine for the 1980s and 1990s, which approximates American-NATO models, gives a partial answer to the question of what objectives a sharp buildup in the nuclear-missile might and the modernization of the armed forces pursue. The previous doctrine, elaborated after France left the NATO military organization, limited the use of nuclear weapons strictly to defending its own territory. For the doctrine being reviewed, it is now a matter of expanding the notion of France's "vital interests" beyond the limits of the country's territory.[48]

Whatever the ambiguities surrounding the question of the independence of French nuclear forces, French conventional forces deployed in Europe—even those on French soil—are seen to be an integral component of NATO. France's political commitment to the Western Alliance is perceived to have led to genuine cooperation between French conventional forces and those of NATO. For example, Marshal N.V. Ogarkov has stated that "in the past few years France has been increasingly departing from the known line of General de Gaulle, increasing in every way practical military cooperation with NATO, up to and including coordinating the operational plans of its armed forces with NATO's general plans."[49]

From the beginning of its independent policy, France retained close ties with NATO, continuing its work in several NATO military committees. France's military relationship with NATO deepened further in the 1970s and 1980s. Most notably, the French armed forces have participated more frequently in combined exercises with NATO forces. For Soviet analysts there is a close link between the cooperation in military exercises and the increased salience of the Western Alliance to French security policy. Captain I. Volodin has articulated this linkage in the following way: "The statement by the country's military-political leadership about its loyalty to its North Atlantic bloc allies, which is regarded by Western military specialists as direct support for NATO's military interests, is indicative of the increased participation of France's navy in recent years in the combat and operational training of the NATO Joint Armed Forces."[50]

In addition to participating in joint military exercises, France is taking a leading role in encouraging West European nations to promote indigenous armaments production. Although France is not a member of the Eurogroup component of NATO, France has become a key member of the Eurogroup committee dealing with armaments production. France participates in order to promote French arms sales, but the effect is to promote greater West

European production of conventional arms, a trend not favored by the Soviet Union.[51]

French conventional forces do, however, have an important capability for independent action, especially outside Europe. Notably, the navy provides French leaders with an ability to project power and to exercise influence in Africa and the Middle East. But even French military actions outside Europe have a connection with the Alliance. Of the European NATO allies of the United States, only France and Great Britain have the real capability to involve themselves in Alliance military operations outside Europe. Soviet analysts have underscored that French actions in Chad and Lebanon have "encouraged" the Americans to seek greater Alliance cooperation in military actions beyond NATO's boundaries.[52]

Soviet analysts tend to assume that the Americans' insistence on formal commitments by France to participate in NATO actions within and outside Europe will continue to create serious tensions with the French. The French desire to stay out of the integrated military command and the American desire to have France rejoin the command will keep providing the Soviets with the hope that intra-Alliance tensions will continue to inhibit Alliance coordination.[53]

In summary, Soviet analysts view French military independence as best understood in terms of deploying and trying to protect and control an independent nuclear force. French conventional forces, however, are considered an integral component of NATO forces in Europe. Perceived changes in recent French security policy are seen simply to reinforce the basic judgment that French conventional forces are highly involved in NATO planning and operations in the European theater. French conventional forces outside Europe are seen as a major means of exercising influence in the Third World and hence a means of satisfying French aspirations for a meaningful role in global affairs. Soviet analysts do, however, note a high degree of cooperation between the American and French navies outside Europe and have been concerned that operational cooperation might be transformed into sustained political cooperation outside Europe.

Based on Soviet perceptions of the French force posture, it is likely that Soviet military planners have reached the following judgments about French behavior in a future war in Europe. First, the probability that French forces will be involved in NATO's conventional operations is high enough that Soviet military planners must prudently plan to operate against them. Second, France will most likely use nuclear weapons to defend its territory; other uses remain much more problematic. Third, if the Soviets begin a full-scale war in Europe, they might well be willing to run the risk of trying to intimidate the French into not using their nuclear weapons during Soviet conventional operations. Soviet conventional forces would probably have a primary mission of degrading not only NATO nuclear weapons in the European theater but French nuclear forces as well.[54] Soviet targets probably would include French ballistic missile submarines.[55]

The Political and Military Impact of French Security Policy

For Soviet analysts, the significance of French security policy lies in its impact on American and West European policy and in the subsequent positive or negative consequences for Soviet security policy.

The major political effect of French policy has been to stimulate the Europeanist tendency in Western Europe. A. I. Utkin has identified Europeanism as "an ideology of isolating Western Europe, of forming the West European alliance as an autonomous center in the world arena." Utkin has emphasized that Europeanism has worked at cross-purposes to Atlanticism; its focus on greater independence from the United States has resulted in an anti-American element to Europeanism.[56]

France has played a critical role in challenging American hegemony and in encouraging the emergence of Europeanism as a practical ideology guiding the formation of a more independent West European power center. According to V. P. Lukin, "there is scarcely any doubt that it was the Gaullist trend that played the decisive role in forcing Atlanticism into the background . . . and in breaking through the political surface of the Europeanist attitude and practical initiatives."[57]

Europeanism as a trend and France's critical role in promoting it are seen by Soviet analysts as a two-edged sword. Europeanism is a positive trend to the extent that Atlanticism or American hegemony is undercut, but a negative trend to the extent that a West European power center emerges as a result. French policy and the Europeanist tendency with which it is associated have undercut American hegemony. France's withdrawal from the integrated military command means to the Soviets that French compliance with U.S. policy can never be assumed by Washington. The French can act as useful interlocutors between the Soviets and the Americans or between the Soviets and other West European states. From the Soviet point of view, it is better for France to be a "perturbator" in West-West relations than a consistent supporter of American positions.[58]

The withdrawal of France from the integrated military command has also had important consequences for U.S. military policy. First, the Americans have lost the automatic use of French territory in a war. Second, the French have consistently rejected the American doctrine of limited nuclear war. According to Soviet analysts, the Americans plan to fight various forms of limited nuclear war, whereas the French plan for an all-out nuclear war.[59] The clear implication is that French nuclear forces seriously complicate American plans for exercising escalation control over Western nuclear forces. Third, the French example has encouraged other Western states, such as Spain, in the direction of only partial involvement in American war-fighting plans.

Although the French policy of independence has yielded benefits from the Soviets' point of view, they nevertheless perceive France's role in strengthening a West European power center as largely negative. To gain some understanding of how Soviet analysts generally treat the nature, prospects, and challenges of the Europeanist tendency, the argument of one Soviet analyst is presented here. The work analyzed is a chapter by V. P. Lukin on the emergence of the West European power center in his book *"Power Centers": Conceptions and Reality.*

Lukin first discusses the emergence of a West European power center in the economic sphere, where there have been objective shifts in the correlation of forces between Western Europe and the United States.[60] The narrowing of the economic gap between the two power centers has contributed to a weakening of U.S. political domination of Western Europe.[61]

West European economic power, however, has provided only the prerequisite for the emergence of West European political power. A major force retarding the political emergence of Europe has been the continued military domination of Western Europe by the United States. In Lukin's view, the political emergence of Europe is inseparable from its emergence as a military power, even if firmly allied with the United States. Stimulating the emergence of West European military power is the growing doubt about the American nuclear guarantee. The French heresy of the mid-1960s in casting aspersions on this guarantee has become West European orthodoxy.[62]

At the core of a West European military power center would be greater military cooperation among France, West Germany, and Britain. The Mitterrand and Chirac administrations have been viewed as major proponents of such a course of action for Western Europe. Lukin cites official French statements and the government's attempt to "set up the activities of the Western European Union" as proof of France's determination to adhere to this policy.[63]

Lukin is well aware of the numerous difficulties that confront the emergence of an independent West European military power. First, Atlanticist tendencies remain strong in many parts of Western Europe. "Within the ruling circles of the West European states (especially in the FRG, England, and Italy) sentiments continue to remain strong that favor the improvement of American–West European relations in the military sphere in the Atlanticist mode."[64] Second, tensions among West European states weaken the trend toward greater strategic cooperation, particularly between France and Great

Britain: "Sharp political contradictions between France and England have, until now, placed nuclear cooperation between them outside the limits of what is politically possible."[65] It is also difficult for the Europeans to resolve the problem of how to integrate West Germany into a West European military organization. Third, France's nationalistic tendencies, especially with regard to nuclear policy, tend to impede military integration.[66]

Nonetheless, "in the long term, the Europeanist power-center tendencies of Western Europe's ruling circles will, apparently, gradually gather strength and will be reflected in concrete political actions."[67] The sense that Atlanticism is in irrevocable decline and the relative independence of Europe is on the rise is at the core of Soviet analyses of France in particular and of European security more generally. For Soviet analysts, Western Europe's independence can develop in one of two ways: Europe could become more independent militarily, or it could become more neutralist and embrace demilitarization.[68] There is little doubt which of these two alternatives the Soviets favor. Equally certain is that France is contributing to the emergence of the former.

Thus, the significance of French security policy to Soviet analysts rests largely on its contribution to the Europeanist tendency. Although France's example of independence has been important in stimulating the quest by West Europeans for greater independence from the Americans, France's contribution to security interdependence within the West European power center is more worrisome to the Soviets.

Soviet analysts have noted a number of ways in which the evolution of French policy will decisively affect the emergence of security cooperation within Western Europe. First, to the extent that France develops its cooperation with Great Britain, a true European nuclear force (a Soviet term) could emerge.[69] Second, as the Franco-German security relationship becomes stronger, the Germans are more likely to acquire nuclear weapons, indirectly or directly.[70] Third, to the extent that the major West European powers, perhaps through the WEU cooperate in joint armaments production, the possibility of greater cooperation in deployment of conventional forces is enhanced. Weapons standardization enhances the probability of conventional military cooperation.[71] Fourth, the French example of using arms to gain influence in the Third World might well become the dominant tendency for the West European armaments industry. The Soviets have noted that France is already the second leading arms producer in the capitalist world; because of the significance of arms sales for their own hard currency earnings, the Soviets might be concerned with the growing threat European armament producers pose to Soviet arms markets.[72]

At least some Soviet analysts have posed the possibility that Atlanticism and Europeanism might not be polar opposites. Rather, Europeanism might provide the Americans with more room for maneuver in dealing with contingencies outside Europe. Actions taken by the Mitterrand administration to promote European cooperation and to reduce the anti-American thrust associated with traditional Gaullism are precisely a manifestation of something approaching a worst-case scenario for the Soviets. According to Captain V. Kuzar', "the strengthening of NATO's 'European wing' would free the United States' hands even more for aggressive actions beyond the geographical framework of the bloc's 'zone of responsibility.'" Kuzar' also notes the increased involvement of several NATO countries in "out-of-area" issues, such as the peace-keeping forces in Lebanon and in the Persian Gulf.[73]

Besides contributing to the Europeanist tendency within the Western Alliance, French security policy has affected other areas of significance to the Soviets, such as Soviet arms control diplomacy. France's stance on arms control has been consistently negative from the Soviet point of view, with regard to both U.S.-Soviet bilateral agreements and the noninclusion of French nuclear forces in any realistic forum. Although the Soviets finally abandoned their insistence on including French and British nuclear forces in the INF talks, it should not be anticipated that Soviet negotiators will relinquish all claims to eliminate (or at least substantially reduce) these forces in future arms control arenas. Since the INF agreement, the negative aspects of French disarmament policy have continued as Soviet analysts repeatedly point to the leading role of France—along with Great Britain and the United States—in seeking ways to compensate for abolition of the INF systems.[74] France's intransigence has also impeded Soviet efforts to obstruct modernization of the British and French nuclear forces, which could form the basis for a future European nuclear force.[75] Furthermore, it has encouraged Chinese intransigence and has provided a basis for cooperation between China and France in developing strategic nuclear weapons outside the regulation on U.S.-Soviet arms control agreements.[76]

A second issue of significance to the Soviets is the effect of French security policy on Soviet military strategy. If the Soviets have made conventional warfighting their basic military option, then the importance of France in Soviet calculations should be growing.[77] The French nuclear force, especially as it is modernized, will be of such significance that the Soviets might plan to target it early in a European conflict. The Soviets could conceive of a range of conventional strikes against French land-based nuclear assets and strategic nuclear submarines. It is extremely unlikely that the Soviets will want France to retain the option of escalating to nuclear weapons, thereby complicating or conceivably defeating their conventional war-fighting strategy.

Another dimension of the impact of French security policy on Soviet military strategy relates to the Soviet attention being devoted to the coalitional nature of any future European war. Soviet emphasis is shifting away from simply studying the ability of the Americans to conduct combined NATO operations through the integrated military command and toward the ability of the Alliance to congeal politically in crisis situations. From this standpoint, the absence of France from the integrated military command of NATO is less significant than France's effect on the cohesiveness of the major powers in the Western coalition.

A third issue of significance to the Soviets is the effect of French security policy on Soviet policy in the Third World, especially in the Middle East. In light of the close ties France has developed with the Arab Middle East, the Soviets now have to compete with both the Americans and the French if they are to reduce the Western presence in the region. U.S. reverses or difficulties in the region do not simply redound to Soviet advantage, for often the French, not the Soviets, are the beneficiaries of American difficulties. French arms sales, especially in the aerospace industry, have frequently been made at Soviet expense.[78]

In short, the significance of French security policy for Soviet analysts is its effect on American options and on West European possibilities. The French policy of independence and the nuclear force structure with which it is associated are seen to have been, and to continue to be, critical factors affecting U.S. military policy in Europe. French policy has weakened U.S. conventional options in Europe. It has also complicated the American ability to control escalation in a European war. French independence, however, challenges Soviet interests to the extent that French policy stimulates and cooperates in a process of interdependence among the major West European states. Insofar as Europe is better able to defend itself, the Americans are left in a more flexible position to act globally. And as the French nuclear forces continue to be modernized, they will become a more significant problem for Soviet military strategy.

Notes

1 Research for this chapter was completed in December 1988.
2 See, for example, the point of view expressed by O. N. Bykov in *Zapadnaya Evropa i SShA: ocherk politicheskikh vzaimootnoshenii [Western Europe and the U.S.: An Essay of Political Interrelations]* (Moscow: Mysl', 1968), chap. 1.
3 See, for example, V. Baranovskii, "The Azimuths of France's Foreign Policy," *MEMO*, p. 154.
4 G. P. Chernikov and D. A. Chernikova, *Storonniki i protivniki franko-sovetskogo sotrudnichestva [Proponents and Opponents of Franco-Soviet Collaboration]* (Moscow: Mezhdunarodnye otnosheniya, 1971), p. 39.
5 A. Vladimirov, "The Pernicious Consequences of 'Nuclear Balckmail'," *MZ*, 7 (1963), pp. 138–139.

6   Zhan Kanapa, "France: Militarization of the Economy and Politics," *MZ*, 9 (1963), p. 49.
7   For an interesting treatment of the tensions in de Gaulle's European policy, see Yu. I. Rubinskii, ed., *Frantsiya [France]* (Moscow: Mysl', 1973), pp. 418–426. For a statement of the conflict within French ruling circles between aspirations toward detente and West European integration, see V. Rakhmaninov, "Soviet-French Relations and European Security," *IA*, 11 (1970), p. 33.
8   V. A. Zorin, "Continuity and Change in France's Foreign Policy," in L. S. Voronkov, G. A. Vorontsov, and V. A. Zorin, eds., *Vneshnyaya politika kapitalisticheskikh stran [Foreign Policies of the Capitalist Nations]* (Moscow: Mezhdunarodnye otosheniya, 1983), p. 109.
9   See, for example, T. Vladimirov, "Factor of Peace and Stability in Europe," *IA*, 8 (1970), p. 96, and Yu. Vladimirov, "Soviet-French Cooperation: Steady Progress," *IA*, 8 (1971), pp. 69–71.
10 V. P. Slavenov, *Vneshnyaya politika Frantsii, 1974–1981 [France's Foreign Policy, 1974–1981]* (Moscow: Mezhdunarodnye ontosheniya, 1981), pp. 112–113.
11 T. I. Sulitskaya, *Kitai i Frantsiya, 1949–1981 [China and France, 1949–1981]* (Moscow: Nauka, 1983), p. 135.
12 A. Kovalev, "Certain Aspects of French Military-Strategic Doctrine," *MEMO*, 4 (1978), p. 120. See also Col. G. Vasil'ev, "France's Defense Expenditures for 1979," *ZVO*, 5 (1979), pp. 24–25.
13 A. I. Utkin, "The Current Phase in American-French Relations," *SShA*, 11 (1975), pp. 26–36.
14 S. Vorontsova, "The U.S. and France: Spheres of Cooperation and Contradiction," *MEMO*, 11 (1980), pp. 61–62.
15 Yu. P. Davydov, "Washington's Course toward Tension and Western Europe," *SShA*, 10 (1980), p. 32.
16 Yu. P. Davydov, "U.S.-Western Europe and the Limits of Compromise," *SShA*, 6 (1975), p. 35.
17 Davydov, *SShA*, 10 (1980), pp. 33–34.
18 A. Kudryavtsev, "France—Political Shifts," *MEMO*, 10 (1981), pp. 103–112.
19 A. Kudryavtsev, "France in the World—A Year Later," *MEMO*, 10 (1982), p. 57.
20 Ibid., p. 58.
21 V. Pustov, "A Growing Threat: Nuclear Forces," *KZ*, January 31, 1984, p. 3.
22 For a significant overview of the Mitterrand administration's military policy, see Col. Yu. Erashov, "The Dangerous Evolution of France's Military Policy," *ZVO*, 9 (1983), pp. 13–17.
23 Col. Yu. Erashov, "Western Europe in the Vise of NATO," *KVS*, 18 (1983), p. 77.
24 N. Leonidov, "What Is Happening with France's Military Policy," *KZ*, April 19, 1983, p. 4.
25 Yu. Yakhontov, *Pravda*, June 8, 1983, p. 5; A. Grigoryants, *Izvestiya*, March 4, 1984, p. 4.
26 Maj. V. Nikanorov, "Indulging Militarism," *KZ*, May 18, 1984, p. 3.
27 Yu. Buskin, "Jaguars Approach the Target," *Moscow News*, December 18, 1983, p. 7.
28 For a Soviet statement that identifies 1976 as the turning point, see P. Cherkasov, "Tendencies of Atlanticism," *MEMO*, 1 (1982), p. 147.
29 G. Dadyants in *Sotsialisticheskaya industriya*, July 9, 1983, p. 3. See also, for an excellent summary of Soviet thinking about the Mitterrand administration's security policy in the early 1980s, Z. Arsen'ev, "France and the Treachery of Cold Winds," *Sovetskaya Rossiya*, April 18, 1984, p. 7.
30 V. S. Mikheev, "Washington-Paris: The Current Stage of Relations," *SShA*, 2 (1983), p. 21.

31 T. V. Kobushko, "Washington against the Gas Pipeline Agreement," *SShA*, 12 (1982), pp. 46–55.
32 This section of the Mitterrand years' analysis was researched and written by Daniel Peris and Susan Clark.
33 S. Borisov, "The Balance Is Positive," *Novoe vremya*, 12 (1985), p. 17.
34 "M. S. Gorbachev's Speech," *Pravda*, July 8, 1986, p. 2.
35 V. Bol'shakov, "F. Mitterrand Reelected President," *Pravda*, May 10, 1988.
36 Mitterrand and Gorbachev met in June 1984, October 1985, July 1986, and November 1988. Also, Chirac met with Gorbachev in May 1987.
37 A. Kudryavtsev, "France and the American 'Star Wars' Program," *MEMO*, 10 (1985), p. 68.
38 Ibid., p. 69.
39 Ibid., p. 71.
40 Davydov, *SShA*, 10 (1980), p. 31.
41 See, for example, S. Rudas, "France's Navy: Prospects for Development," *Morskoi sbornik*, 1 (1977), p. 96; Capt. I. Volodin, "France's Naval Aviation," *ZVO*, 7 (1981), p. 65; Lt. Col. A. Pavlov, "France's Air Force," *ZVO*, 4 (1982), p. 44.
42 Col. N. Frolov, "France's Ground Troops," *ZVO*, 11 (1980), pp. 25–32; Col. N. Frolov, "France's Armed Forces," *ZVO*, 1 (1985), p. 13; N. K. Glazunov and N. S. Nikitin, *Operatsiya i boi [Operations and Engagements]*, 3d ed. (Moscow: Voenizdat, 1983), pp. 265–269.
43 I. A. Koloskov, *Vneshnyaya politika pyatoi respubliki: Evolyutsiya osnovnykh napravlenii i tendentsii, 1958–1972 [The Foreign Policy of the Fifth Republic: The Evolution of Basic Directions and Tendencies, 1958–1972]* (Moscow: Nauka, 1976), p. 287. See also Rubinskii, ed., *Frantsiya*, pp. 359–361; Mikheev, *SShA*, 1 (1986), p. 28.
44 Rubinskii, ed., *Frantsiya*, p. 349.
45 E. A. Arsen'ev, *Frantsiya: Problemy i politika [France: Problems and Politics]* (Moscow: Politizdat, 1978), p. 248.
46 See, for example, Capt. I. Volodin, "France's Naval Forces," *ZVO*, 9 (1980), p. 59; Capt. 2nd Rank S. Grechin, "French Nuclear-Missile Submarines," *ZVO*, 4 (1985), p. 63.
47 Capt. 1st Rank A. Karemov and Col. (Ret.) G. Semin, "The Military Doctrines of the Main European NATO Countries," *ZVO*, 7 (1983), pp. 17–18.
48 I. Shchedrov, "Myths and 'Reality'," *Pravda*, June 1, 1983, p. 4.
49 MSU N. V. Ogarkov, "Peace Must Be Defended Reliably," *Soviet Military Review*, 12 (supplement) (1983), p. 17. For other sources that view French conventional forces as clearly integrated into NATO, see especially the map in Col. A. Alekseev, "The Airfield Network of the Main NATO Countries in Europe," *ZVO*, 1 (1984), p. 65; Mikheev, *SShA*, 1 (1986); Frolov, *ZVO*, 1 (1985), p. 12.
50 Volodin, *ZVO*, 9 (1980), p. 59.
51 See V. Leushkanov, "Atlantic Arms Generator," *MEMO*, 9 (1978), pp. 113–115; V. Naumov and Yu. Pokataev, "NATO's Eurogroup," *Voennye znaniya*, 4 (1984), p. 31.
52 See, for example, Zorin in *Vneshnyaya politika*, pp. 129–130; "Europe in the World Politics of the Eighties," *MEMO*, 2 (1984), p. 9.
53 This judgment is reflected throughout S. B. Vorontsov, *SShA i Frantsiya: sopernichestvo i partnerstvo [U.S. and France: Rivalry and Partnership]* (Moscow: Mezhdunarodnye otnosheniya, 1983).
54 Phillip Petersen and Maj. John Hines, *The Soviet Conventional Offensive in Europe* (Washington, DC: Defense Intelligence Agency, 1983).
55 I. Kuz'min, "Submarine Noise and Anti-Submarine Warfare," *Morskoi sbornik*, 9 (1982), pp. 67–72. Kuz'min included French SSBNs in the list of Western threats to the Soviet Union that required an enhancement of Soviet antisubmarine warfare capabilities.

56 A. I. Utkin, *Doktriny atlantizma i evropeiskaya integratsiya [The Doctrines of Atlanticism and European Integration]* (Moscow: Nauka, 1979), p. 73.
57 V.P. Lukin, *"Tsentry sily": Kontseptsii i real'nost' ["Power Centers": Conceptions and Reality]* (Moscow: Mezhdunarodnye otnosheniya, 1983), pp. 66–67.
58 On the significance of the perturbator role, see Vorontsov, *SShA i Frantsiya*, especially pp. 83–89 and 113–121.
59 Mikheev, *SShA*, 2 (1983), p. 21.
60 For a fuller analysis of Soviet views on West-West conflict, see Erik P. Hoffmann and Robbin F. Laird, *The "Scientific-Technological Revolution" and Soviet Foreign Policy* (Elmsford, NY: Pergamon Press, 1983), chap. 2.
61 Lukin, *"Tsentry sily,"* p. 69.
62 Ibid., p. 82.
63 Ibid., pp. 86–87.
64 Ibid., p. 87.
65 Ibid.; see also V. Pustov, "Paris and NATO," *KZ*, March 12, 1988, p. 5.
66 Ibid.
67 Ibid., p. 89.
68 See V. S. Shein, "U.S.-EEC: A Bundle of Contradictions," *SShA*, 1 (1973), p. 63.
69 For an excellent overview of France's increased ties with NATO, see Pustov, *KZ*, March 12, 1988. For the fullest statement of the European nuclear force concept, see V. F. Davydov, "Discussion of the European Nuclear Forces," *SShA*, 3 (1976), pp. 28–38.
70 See, for example, V. Mikhnovich, "Calculations and Miscalculations," *KZ*, June 14, 1984, p. 3; Pustov, *KZ*, March 12, 1988.
71 For a representative Soviet perspective on the Western European Union, see D. Proektor, "A Carte-Blanche to Rearm," *Novoe vremya*, 19 (1984), pp. 20–21. See also Col. M. Ponomarev, "At Washington's Urging," *KZ*, March 11, 1984, p. 3.
72 I. Shchedrov, "France and the Pershings," *Pravda*, June 15, 1984, p. 4. See also the argument contained in *Soviet Arms Trade with the Non-Communist Third World in the 1970s and 1980s* (Washington, DC: Wharton Econometrics Forecasting Associates, 1983).
73 V. Kuzar', "NATO—An Alliance in the Name of Aggression," *KZ*, April, 8, 1984, p. 3.
74 See, for example, Pustov, *KZ*, March 12, 1988.
75 Pustov, *KZ*, January 31, 1984.
76 Sulitskaya, *Kitai i Frantsiya*, chap. 3.
77 On the shift in Soviet strategy to the conventional option, see Robbin F. Laird and Dale R. Herspring, *The Soviet Union and Strategic Arms* (Boulder, CO: Westview Press, 1984), chap. 1.
78 See, for example, David B. Ottaway, "U.S. and France Compete in Persian Gulf," *Washington Post*, April 25, 1984, and *Soviet Arms Trade*.

6
Soviet Perspectives on British Security Policy[1]

Robbin F. Laird
Susan L. Clark
Institute for Defense Analyses

This chapter constructs an assessment of how the Soviets view the evolution, nature, and significance of British security policy. Because of the centrality of Great Britain to both the United States and to the NATO coalition, the Soviets pay considerable attention to British security policy. Following an overview of the Soviet perspective of the evolution of British security policy in recent years, this chapter then looks at the Soviet treatment of the main elements of the British force structure. The final section examines in some detail Soviet analyses of three major elements of British security policy: the British-American relationship, the British approach to the Europeanization of the Western Alliance, and the nature of the British security debate and its implications for the Soviet Union.

The Evolution of British Security Policy

Developments during the 1970s

While World War II demonstrated the potential for cooperation between the Soviet Union and Great Britain, the state of postwar relations between these two nations has been anything but stable. Soviet analyses of relations between the Soviet Union and Great Britain during the 1970s concentrates on such issues as the U.S.-British special relationship, the differences and similarities between Labour and Conservative party policies, Europeanism versus Atlanticism in Great Britain's foreign policy, and detente. The Conservatives held power during the first part of the decade (until 1974), and the Labour party held office in the second half (until 1979). Although fluctuations in Soviet-British relations generally tended to parallel the changes in Great Britain's political leadership, the Soviets were not as successful as they had perhaps expected to be in improving bilateral relations during the Labour party's tenure in office.

During this decade, Soviet analysts perceived the movement in British policy away from Atlanticist trends and toward a greater European orientation to be "the most significant shift in the evolution of British foreign policy over the past ten years."[2] Under Prime Minister Edward Heath's Conservative government, British thinking reasoned that Europe's only hope for playing an important role in the world lay in its unification. Great Britain wanted to accomplish this by uniting nation states, however, not through supranational organizations. The Soviets point out that in the diplomatic, economic, and military spheres, Heath sought to do just this by intensifying Great Britain's West European focus and deemphasizing its relationship with the United States.[3]

But when Harold Wilson's Labour government came to power in 1974, it rejected the Europeanist tendencies that Heath had espoused. In terms of defense issues, Labour continued to be less committed to defense and therefore fell back on Great Britain's traditional Atlanticist mechanisms, which allowed the nation to rely more on American contributions than on its own. In explaining this development, one Soviet analyst stresses that this difference between the Conservatives and Labour is determined more by the situational context than by any difference in their foreign policies:

> Essentially there are no differences in principle between the two approaches—Labour and Conservative.... The foreign policy platforms of the two parties ... are determined by the specific situation. The Conservatives have listed greatly toward a European orientation which has been dictated by the needs of the moment: England drawing nearer to the EEC and the presence of sharp contradictions with the United States. The Labourites, on the other hand, have paid greater attention to Atlanticism precisely during a period of temporary compromise between the U.S. and Western Europe.[4]

A book by S. P. Madzoevskii and E. S. Khesin identifies three distinct phases in the U.S.-UK special relationship: before the worldwide economic crisis of 1974–1975, the second half of the seventies until 1979, and since the Conservatives came to power under Prime Minister Margaret Thatcher in 1979. According to the authors, although the transition from the first stage to the second was partially attributable to the coming into power of a more pro-American Labour government in 1974, the change was actually due more to economic considerations, especially the oil crisis.[5] The second stage, during Labour's rule, witnessed a revival of the special relationship, both in Great Britain's stated policy of trying to serve as the bridge between Europe and the United States and in continued Anglo-American nuclear collaboration.[6] In sum, Great Britain was seeking to preserve the special link it shared with the United States within the NATO framework, with the nuclear partnership remaining the core of this relationship. In addition, although Great Britain had played a key role in the creation of NATO's Eurogroup, a fact that reflected its Eurocentrist tendencies, this action did not detract from the Anglo-American relationship since the Eurogroup was designed to make the NATO alliance more viable and vital.[7]

Another topic that has remained central to Soviet writings on Great Britain is the role of the two main political parties in Soviet-British relations. Even though the Labour governments of the 1960s and 1970s certainly did not always meet Soviet interests, there is no doubt that Soviet analysts perceived a clear advantage in dealing with a Labour government over a Conservative one. Moreover, during the early 1970s, while Labour was in opposition, the party "came out more decisively in support of a number of the USSR's proposals in the area of disarmament and strengthening European security and universal peace."[8] Labour's return to power in 1974 continued many of the general trends begun while it was in opposition. Wilson's government encouraged the thawing of current relations and the development of more constructive relations with the East, as evidenced in his own trip to Moscow in 1975 for high-level talks with the Soviet leadership.[9]

In general, Soviet analyses of the Labour party on nondefense issues were relatively positive. In contrast, in terms of Labour's defense policies, the Soviets recognized that campaign promises are not necessarily translated into government actions. For example, whereas in its preelection manifesto of 1974 the Labour party declared its intention "to limit the arms race [and] to raise the question of abolishing U.S. bases for missile submarines at Holy Loch 'as a first step' at the arms reduction talks," the program it published after assuming office looked quite different. The latter underscored the priority of NATO to Great Britain's foreign policy and does not mention plans to refuse to create a new generation of nuclear weapons or to abolish U.S. bases on British soil.[10]

In his synopsis of the Labour years, 1974–1979, V. A. Ryzhikov emphasizes that the government continued a firm commitment to NATO and "to all intents and purposes, . . . completely supported U.S. military-policy actions."[11] An increased representation of left-of-center Labour forces did make possible some improvement in Soviet-British trade, economic, scientific-technological, and cultural ties.[12] However, under Prime Minister James Callaghan "in 1978 and the beginning of 1979, the inconsistency, instability, contradictoriness, and even anti-Sovietism in Labour's policies toward the USSR intensified."[13] At this point, issues of Soviet human rights abuses, the indivisibility of detente from other international events, and Soviet policies in Third World countries all contributed to a deterioration in Great Britain's relations with the Soviet Union. Ryzhikov concludes by cautioning that there existed "an undoubted gap between the words and deeds of the right-wing Labourites."[14]

Thatcher, Gorbachev, and the 1980s

Not surprisingly, the election of Margaret Thatcher's Conservative government to office in 1979 brought about a deterioration in Soviet-British relations. Madzoevskii and Khesin argue that the common anti-Soviet focus of the governments of both Thatcher and Ronald Reagan contributed to a rapprochement in Anglo-American relations, a development further facilitated by the prime minister's strong commitment to Atlanticism and to the maintenance of the special relationship.[15] In addition, the 1980 agreement signed by President Jimmy Carter and Prime Minister Thatcher for the purchase of the Trident system signified a firm commitment to continued nuclear collaboration, a fact of obvious concern to the Soviets.

Unlike some of Great Britain's previous governments, that of Thatcher has stressed the importance of European cooperation as a means of strengthening the Alliance but not as a means of opposing American policies. According to A. V. Golubev, "strengthening the alliance" means "first and foremost, consolidating 'Atlantic solidarity,' i.e., the readiness of the West Europeans to support U.S. policy."[16] *Pravda* correspondent A. Maslennikov underscored another manifestation of this solidarity, arguing that the United Kingdom has no disarmament policy of its own. Here he quotes Secretary of Foreign Affairs Geoffrey Howe, who cautioned that the USSR would find no serious differences between Great Britain and the United States on arms control positions.[17]

A graphic example of the Atlanticist versus Europeanist approach in British politics emerged during the Westland affair of December 1985, which resulted in the resignation of Secretary of Defence Michael Heseltine. While Heseltine advocated a Europeanist solution to bail out the failing Westland helicopter corporation, Prime Minister Thatcher opted for the American solution. Soviet analysts were aware of the divisiveness of the issue and of its implications for future policies. With the victory of the pro-American solution and the resulting consequence of American penetration into Great Britain's helicopter industry, Maslennikov concludes that "those in British ruling circles cannot help but see that too close and clearly unequal relations with the American partner have raised the threat of sacrificing England's last remnant of independence and authority in international affairs."[18]

While Thatcher's early years in office did not augur well for Soviet-British relations, changes in the Labour party did hold a certain appeal for the Soviets. Soviet writers noted a significant shift in Labour's policies after the beginning of the 1980s. Specifically, at the turn of the decade the positions of the Conservative and Labour parties on nuclear modernization were "essentially one and the same."[19] When the Conservatives announced their decision to purchase Trident I in 1980, Labour's reaction, with the right-of-center leaders still in control, was termed "lethargic and indeterminate."[20] However, the subsequent decision to purchase Trident II instead, with its enhanced capabilities, helped precipitate a surge in the peace movement in Great Britain, on which Labour tried to capitalize. Since the beginning of the 1980s—and especially since Neil Kinnock became the party's head—Labour's defense positions became "considerably more radical,"[21] as evidenced in its adoption of a new defense document in July 1984 calling for universal nuclear disarmament, a no-first-use policy in NATO, a nonnuclear zone in Europe, the renunciation of both Trident and Polaris, and the removal of U.S. nuclear weapons and bases from British soil.

But however much the Soviets may welcome the Labour party's current defense platform, they also realize the continuing difficulties this party faces in trying to get elected. The party is badly factionalized, and it is widely recognized that its defense policy does not appeal to the British public. In fact, during the 1988 annual Labour party conference, Kinnock urged his party to adopt a less radical defense posture (incorporating unilateral, bilateral, and multilateral nuclear disarmament), but he was voted down when the party's advocates of unilateral nuclear disarmament prevailed. To all appearances, this issue will continue to divide the party for the foreseeable future, a fact the Soviets (reluctantly) recognize.

The second half of the 1980s not only witnessed the Labour party's persisting problems in appealing to the electorate, but also revealed indications of a certain improvement in governmental relations between Great Britain and the Soviet Union. At the beginning of her tenure in office, Thatcher cut back ties with the USSR, suspended visits and meetings, and supported some economic sanctions against the USSR. Beginning in 1984, Soviet analysts noted a slight increase in economic cooperation,[22] and certainly Mikhail Gorbachev's advent to power signified a greater willingness to "develop a political dialogue."[23] This dialogue has, indeed, been reestablished in the form of several high-level visits between the two countries. Gorbachev visited London in 1985 and 1987, and Foreign Minister Eduard Shevardnadze visited in July 1986. In turn, Foreign Minister Geoffrey Howe traveled to Moscow in 1987 and 1988, and Thatcher paid a visit in the spring of 1987. The timing of

Thatcher's visit was, in fact, quite important. It was relatively clear that she would win reelection in June 1987; the Soviets accepted this likelihood and used her visit to lay the foundation for improving relations.

Soviet coverage of the Moscow visit emphasized Thatcher's importance as a European leader.[24] Her visit came amid Soviet criticism of the European powers' negative reaction to the Reykjavik summit. Soviet analyses of European reactions emphasized that the conservative governments of Great Britain, France, and West Germany wished to stress nuclear deterrence rather than the elimination of nuclear weapons.[25] Thatcher was perceived to play an important role in shaping this European reaction to Reykjavik. Gorbachev hoped to be able to make her more supportive of the superpowers' efforts to reduce nuclear weapons, in part by dropping the Soviet demand for the inclusion of British and French nuclear weapons in the proposed INF agreement, which he presented as an important Soviet concession.

As a precondition of the Thatcher visit, the Soviets agreed to allow the prime minister to present her views directly to the Soviet public via a televised interview. In this interview, Thatcher made explicit her views on nuclear deterrence.[26] She also informed the Soviet public that it was the Soviet Union, not the West, that had first deployed intermediate-range nuclear force (INF) weapons. Her presentation may have provided the Soviet public with a better understanding of the difficulties the general secretary was having in convincing the conservative European governments to go along with the proposed INF agreement.

Gorbachev's own brief trip to London in December 1987, on the way to signing the INF agreement with President Ronald Reagan in Washington, again underscored the Soviet leadership's appreciation of Thatcher's role as a world leader. The object of extensive Soviet press coverage—including positive assessments of the state of Soviet-British bilateral relations—Gorbachev's visit played on Britain's role in influencing U.S.-Soviet relations and East-West relations more generally. For Thatcher's part, she attributed the improvement in their bilateral relations to *perestroika* and evolving international cooperation that made the INF agreement possible.[27]

During Howe's visit to Moscow in February 1988, Shevardnadze again displayed the more refined and pragmatic diplomatic style espoused under Gorbachev. Shevardnadze recognized American and British traditions and ties while also advocating closer Soviet-British cooperation: "The Soviet Union is far from being an advocate of upsetting Europe's traditional ties across the Atlantic, and does not encroach upon the traditional Anglo-Saxon component of relations between Britain and the United States. Our efforts are aimed at overcoming the division of Europe, and at furnishing and strengthening the common European home."[28] Howe's response was certainly cautious, if not critical. While recognizing in theory the positive consequences of *perestroika* for East-West relations, Howe also emphasized that in practice, few concrete measures have yet been seen, especially in the realm of Soviet foreign policy.

Still, compared to the state of governmental relations between the two countries earlier in the decade, these visits do indicate a desire and willingness on both sides to cooperate more effectively. It should not be expected that the two countries could overcome their fundamental differences on certain key issues. Rather, what seems to be emerging are an acceptance of each country's current realities and ideologies and a determination to develop better working relations within this context. For its part, the Soviet leadership further recognizes that it will have to deal with conservative governments in the key European states for some time to come. Additionally, the Soviets must seek to influence the ongoing process of cooperation among these conservative governments on security matters. Concern about the contribution the British might make to European security cooperation is reflected throughout Soviet literature on Great Britain and is discussed more fully in the third section of this chapter.

The British Force Structure and Military Strategy

Great Britain possesses a diversified force structure consisting of naval forces capable of limited global power projection and an army and an air force postured for operations in the European theater. It also deploys nuclear weapons with all its services. British military strategy relies on the potential use of nuclear weapons as well as joint conventional operations with other member states of NATO to protect national territory and military assets. The British concept of national defense relies heavily on joint operations with its allies to protect British territory. As Captain Karemov and Colonel Semin comment, "The combat use of general purpose forces is anticipated only as part of NATO's joint armed forces, above all in the Central European and Northern European TVDs [theaters of military operations], while the navy would be employed in the Eastern Atlantic and in the English Channel area."[29]

One of the most comprehensive statements regarding British military doctrine, which reveals the Soviet perception of the British mix of nuclear and conventional deterrence, has been made by Karemov and Semin.

> In terms of the scope of military action and the means of destruction used, Great Britain's doctrine distinguishes the following types of wars: a general nuclear war and a limited (including a local) one.
> The doctrine recognizes the unlimited use of nuclear weapons in a general war in the form of a nuclear attack, and the possibility of conducting

limited wars not only outside the NATO zone, but also in Central Europe. At the same time, British specialists react in a very restrained manner to the idea advanced by the White House of unleashing a "limited" nuclear war in Europe, seeing in this an increased threat to the British Isles.

It is believed that a war in Europe with the unlimited use of nuclear weapons can begin by a surprise attack or after a short period of increased tensions as well as through the use of only conventional means of destruction at the beginning of the conflict, then tactical nuclear weapons, and subsequently strategic ones.

In the area of the armed forces' development, [British] military doctrine requires compact, mobile and well-balanced troops and naval forces which would be capable of ensuring the attainment of political objectives and the execution of strategic tasks in any likely conflicts.[30]

Note the heavy emphasis in this analysis on British reliance on nuclear weapons as well as on a diversified force structure to conduct conventional operations.

The British, like the French, possess a full gamut of forces. The British and French have resisted pressures for a style of European cooperation that would have them specialize in certain areas (such as naval deployments) at the expense of relinquishing involvement in other areas (for example, ground forces deployed in West Germany).[31] A key embodiment of British as well as French national interests in the form of a military strategy has been the requirement to be able to fight along the full spectrum of conflict in the European theater with a full range of forces.

The British, like the French but unlike any other European power, wish to have a flexible force structure in order to operate outside Europe as well. British power projection forces rest on the existence of a powerful navy, out-of-area bases, and mobile ground forces. The Falklands conflict was seen by the Soviets as the example par excellence of the British desire to stay in the power projection game. Captains Marov and Biryusov commented that the Falklands campaign demonstrated that

> the naval forces of Great Britain and other capitalist states both now and in the near future will be a reliable instrument for the demonstration and direct use of military force beyond the boundaries of their territory, including in areas great distances away. An important place in the achievement of British imperialism's expansionist objectives is accorded to forward permanent bases and basing stations near regions of possible conflicts. In this connection, it is emphasized that without the base on Ascension Island, Great Britain would not have been able to conduct combat actions successfully in a region so remote from Britain as the South Atlantic.[32]

British participation, let alone success, in the Falklands campaign would have been impossible without the British Navy. Nonetheless, the campaign also demonstrated some serious deficiencies in the navy, notably the absence of effective airborne early warning systems.[33] The campaign also raised questions about the vulnerability of surface ships to surface-to-ship missiles.[34]

The Royal Navy's main missions, however, are not perceived to be the conduct of campaigns like the one that occurred in the Falklands. Rather, their main missions are to participate in combined operations, primarily with the U.S. Navy, in the European theater. According to Captains Galkin and Grechin, the Royal Navy's main missions in order of priority are "to deliver nuclear-missile strikes against targets in the opponent's territory; to destroy—with tactical nuclear weapons as well—his ship groupings, surface ships, and submarines so as to gain and maintain supremacy in the area of combat action; to render support to ground troops on maritime axes; to conduct amphibious assault operations; to protect sea lines of communication, and to ensure the guarding of the 200-mile fishing zone and oil fields in the North Sea both in peacetime and wartime".[35] Lieutenant Colonel S. Anzherskii phrases the first priority of the navy somewhat differently: "The fundamental combat tasks of Great Britain's Navy are: to deliver nuclear-missile strikes against *vitally* important targets in the opponent's territory. [Emphasis added.]"[36]

This sequence clearly suggests the priority, in Soviet eyes, that the British place on the protection of national territory in their military strategy. The delivery of strategic nuclear strikes is to serve independent national purposes. The priorities also show that the British consider the best way to protect Great Britain from conventional attack is to augment British striking power by cooperation in joint operations with its NATO allies. Galkin and Grechin note that Great Britain's naval forces are especially active in NATO exercises conducted in the Eastern Atlantic and Iberian Basin as well as in the Mediterranean Sea.[37]

The Royal Navy's acquisition program is designed to maintain and even augment its capability in the years ahead. The navy's strategic arm is being augmented by the purchase of the Trident submarine from the United States to replace the Polaris. The British are seeking to maintain the surface navy by acquiring new frigates like the Leander-2 and to enhance the capability of subsurface forces with new attack submarines.[38] Furthermore, British plans to build a third small aircraft carrier for non-European operations and to create a force for operations outside the traditional NATO zone are seen by A. I. Utkin as evidence of a weakening of the "Suez syndrome" (that is, fear of foreign entanglements), particularly after the Falklands war.[39]

Unlike the Royal Navy, the Royal Air Force's (RAF) role is limited to European missions. According to Col. V. Artyomov, the main air force missions in order of priority are to support ground troops and naval forces in the European theater of war and in the Eastern Atlantic; to cover British territory and sea lines of communication from the air; to deliver nuclear strikes against enemy targets in the operational and strategic depths of his defense; to gain and maintain air superiority; to isolate areas of combat action so as to prohibit the transport of reinforcements by the opposing side; and to carry out tactical and strategic transports of troops and military cargo via air both within the theater and to other TVDs.[40]

A later listing of the air force missions suggests a more important role for deep interdiction that may reflect the growth in RAF capabilities associated with the acquisition of the Tornado. Lieutenant Colonel Anzherskii argued that the first priorities of the RAF were "the tasks of destroying major troop groupings and important targets in the opponent's territory with the use of both nuclear and conventional weapons."[41]

The preponderance of these missions requires joint operations with Great Britain's NATO allies. Nonetheless, the Soviets note that Great Britain's role in these joint operations is primarily to protect British territory or British forces deployed in Europe: "The Air Force command in Great Britain..., which is placed under the authority of the NATO command, possesses the forces and means to conduct independent air operations using both conventional and nuclear weapons. It executes tasks both on its own nation's territory and in NATO's ground and naval TVDs, primarily in Central Europe and the Eastern Atlantic."[42] The British way to promote its military power is to induce NATO to conduct joint operations to protect British territory and military assets. Also, RAF forces have the personnel and equipment to conduct independent air operations using both conventional and nuclear weapons.[43]

The RAF is given high priority in the British defense budget. Approximately 40 percent of the equipment budget goes to supporting RAF requirements.[44] A key element of the high cost of the RAF is the current acquisition of Tornado aircraft. The deployment of the Tornado significantly enhances the RAF's striking power. The Soviets believe that the British will use the Tornado, in part, to conduct deep interdiction strikes against Warsaw Pact forces. According to Colonel Shiryaev, "On the whole... with the completion of the transition of the main units and subunits of the British air group in the FRG [Federal Republic of Germany] to Tornado aircraft, its combat potential will grow several times over, particularly in [its capability to] deliver strikes against the opponent's second echelons and reserves."[45]

The RAF plans to replace its Jaguar aircraft with 250 European Fighter Aircraft (EFA), being developed jointly with other West European countries, above all with West Germany. The Jaguar was developed to provide a light, highly maneuverable fighter that could be used to deliver strikes against ground targets, to provide direct air support of the ground forces, and to conduct air reconnaissance.[46] The missions of the EFA will be "to provide direct support for ground troops and to isolate regions of combat action, as well as to conduct air engagements when encountering enemy aircraft while executing tasks of delivering strikes against ground targets."[47]

Similar to the RAF, the British army's role is primarily limited to operations in the European theater. According to Lieutenant Colonel Anzherskii, the British ground forces are "postured to conduct combat action in cooperation with the air force and navy, and also as a part of the NATO joint armed forces in the European theater of war with the use of conventional and nuclear weapons."[48] In an earlier article, Anzherskii noted that the main tasks of the regular army are "to conduct combat action as part of the NATO troops, to protect Great Britain's interests beyond its borders, and to prepare military cadres for mobilized deployment of the ground troops in the event of war."[49]

The army is the most committed of all the British forces to joint operations with Great Britain's NATO allies. With only a residual territorial defense role for the army, the Soviets perceive the British army in Germany (BAOR) and the army's force billeted in the United Kingdom to reinforce the BAOR as integral components of the NATO coalition's capability to conduct joint operations in West Germany. These operations are primarily perceived to be conventional in character. The expectation seems to be that the RAF or Royal Navy would be the arms of the British forces most likely to deliver nuclear strikes. Based on Soviet writings, this expectation appears to rest on the following assumptions:

- The BAOR possesses only minimal nuclear warheads.
- Those warheads are controlled by the United States.
- The BAOR is heavily committed to its conventional role.
- The BAOR as a forward deployed force is vulnerable to enemy strikes and as such may be unable to play both a nuclear and a conventional role.

With respect to the army's modernization program, designed to enhance the service's mobility and firepower, the Soviets have noted that this effort is completely oriented toward conventional modernization, which is another indicator of Soviet expectations that the army would in all probability not initiate nuclear fire in the European theater.

The most significant dimension of the British force structure from the Soviet point of view is the nuclear weapons component. All services possess nuclear weapons, a fact that gives a definite nuclear coloring to British military strategy. In terms of tactical nuclear weapons, the navy has the capability to use nuclear depth bombs against submarines and nuclear-tipped missiles against surface shipping. The RAF has the capability to deliver nuclear strikes against Warsaw Pact forces in Eastern Europe. The army has weapons that can be used on the European battlefield to support offensive operations. The heavy-armored divisions are perceived to have the capability to operate on a nuclear battlefield as well.[50]

The strategic component of British nuclear forces is undergoing a significant upgrading, which is of great interest to the Soviets. The first phase of the upgrading was the replacement of the Polaris A-3 warhead with the Chevaline. Soviet reports differ in their judgments regarding Chevaline's capability. Some claim it is a multiple reentry vehicle (MRV) with two or three warheads, but most claim it is a limited multiple independently targeted reentry vehicle (MIRV) with up to six warheads. In a representative statement, Captain G. Perov asserts that Great Britain's military-political leadership believes that

> the realization of the Chevaline program will significantly increase the striking might of the naval strategic force since just one modernized Polaris A-3 missile is capable of ensuring the destruction of several targets in an area of more than 18,000 square kilometers (a missile with the existing warhead [currently covers] an area of 1300 square kilometers) and it will make it possible to ensure the maintenance of a sufficiently high level of combat effectiveness for the SSBNs [nuclear ballistic submarines] until the beginning of the 1990s.[51]

In fact, one Soviet analyst estimates that Great Britain's current nuclear forces "are capable of wiping out ... up to 20 percent of the Soviet industrial resources and manpower."[52]

The design of the Chevaline has been generated by the British desire to be able to penetrate Soviet defenses around Moscow. Indeed, many Soviet analysts consider Chevaline a significant upgrading in British capability to strike the Soviet Union.[53] Perhaps these analysts are simply reflecting more sensitive and pessimistic Soviet assessments concerning the ability of the Soviet Union to defend against Chevaline.

The main Soviet interest is directed, of course, to the second phase of strategic modernization, namely, Britain's acquisition of the Trident submarine with the D-5 missile. The British are planning to build 4 Trident SSBNs with 16 submarine-launched ballistic missile (SLBM) tubes. Most Soviet analysts estimate that each SLBM will carry 8 warheads.[54] Based on this estimate of warheads per SLBM, once the Trident's deployment is complete, Great Britain will possess 512 highly accurate strategic warheads, which will enable it to attack a wide range of military targets. According to Nikolaev, "The implementation of the program to put Trident-2 missiles into service will mean significant qualitative changes in the British strategic force [by] giving it the potential to deliver a first strike from far away, increasing the quantity of warheads, and significantly raising the yield of the missiles and the performance characteristics of the submarines."[55] Interestingly, unlike Soviet treatments of U.S. Trident deployments, the British are not credited with the desire to attack Soviet intercontinental ballistic missile (ICBM) silos.[56]

There are some hints in the Soviet literature regarding possible British employment strategies. First, British warfighting plans and operations for the Trident will be closely coordinated with the United States. As Captain Perov argues, "The United States ... is carrying out the selection of targets for the British SSBNs."[57] The British will rely on American satellites to provide the accuracy necessary for Trident to have hard-target kill probability. In order for the British to follow a strategy of attacking hard targets, they must have American complicity.

Second, the British might well follow a national target plan. They have reserved the right to use the force for national purposes and, in fact, built it primarily for those purposes. As G. V. Kolosov has noted, "the effectiveness of such [nuclear targeting] coordination still depends, in the final analysis, on political factors and on England's readiness to follow plans outlined jointly with the United States. In order to retain control over the British nuclear forces, the British government stipulated in its agreement with the U.S. the right to withdraw them from NATO and to use them independently if the 'higher national interests of England' were threatened."[58] Kolosov added in another source that Britain's leaders have asserted that the main role of British nuclear forces is to deter some kind of threat of a nuclear attack on England, that is, to use its nuclear weapons to protect national territory.[59]

Third, British strategy currently seems directed at political, military, and industrial targets, in that order. According to Captain Grechin, following the withdrawal of the Vulcan bombers in 1982, nuclear submarines are now "the only strategic means of the nation's armed forces. Their combat use is closely tied in with U.S. and NATO plans to unleash a nuclear war, in which Great Britain is accorded an important role in delivering nuclear-missile strikes against state, military, and industrial targets in the territories of the socialist community countries."[60] Note that the British strategic forces threaten the "socialist community," not just the Soviet Union. This statement would seem to reflect an expectation that the British will target not only Soviet territory, but military forces in Eastern Europe as well.

These three judgments together might suggest the following conclusion to Soviet analysts. If the British and the Americans cooperate, the former might use their Trident force to attack significant Soviet military targets up to and including hard targets. This would probably exclude missile silos, for such an attack seems to make little sense from a British point of view. If the United Kingdom goes alone, it can still hit many key political, economic and military targets but probably not the hardest of military targets, namely silos and command posts. However, Trident lends much more credibility to an overall threat to the socialist community, because the British might contemplate a first strike against Soviet forces operating in Europe.

There is even a suggestion by Kolosov that the British might be thinking in terms of a phased escalation strategy.

> In advocating the build-up of nuclear arms, England's leaders are at the same time trying to follow the United States in elaborating strategic plans. For example, Great Britain's Ministry of Defence Blue Book for 1981 made an attempt to justify the U.S. focus on conducting a counterforce strategy. This document argues that the use of nuclear weapons can be controlled, without resorting to a general war of annihilation. "Escalation," claim the document's authors "is not a self-developing process, but is determined by the decisions of people. Therefore, it is entirely reasonable, even necessary, to elaborate plans which would succeed in providing and implementing the potential making it possible to end the war before it leads to a global catastrophe." Preparations in and of themselves for a nuclear war and, mainly, the attempt to convince the nation's public of the possibility and even desirability of conducting it represent an extremely dangerous and alarming tendency in the formation of British military-political doctrine for the 1980s.[61]

Note that in this comment Kolosov ascribes to the British—not just to the Americans—a belief in escalation control.

In addition to the strategic nuclear modernization program, the Soviets continue to pay attention to tactical nuclear modernization as well. According to Kolosov, "Statements by England's Ministry of Defence especially underscore the importance of the broad-scale outfitting of the air force with British-produced nuclear bombs. The 'advantages' of arming naval aviation airplanes and helicopters with nuclear depth bombs are being examined."[62] Of special interest is the role of the Tornado in delivering nuclear bombs as well. An *Izvestiya* correspondent has estimated that after both the Trident and Tornado modernizations, Great Britain will have the capacity to deploy a total of 1,088 nuclear warheads on both its strategic and tactical systems.[63]

To sum up, the British force structure embodies a military strategy with two key dimensions. First, British strategy continues to rely heavily on the potential use of nuclear weapons in a European war. As Kolosov argues, "British leaders and military commanders, while fully recognizing the advantages that come with placing the latest conventional armaments systems into service, and while continuously carrying out such a modernization, do not consider any kind of reduction in the role of nuclear weapons to be possible."[64] Second, British conventional forces are heavily oriented toward joint operations; for example, according to Soviet calculations, the navy plans to operate up to 85 percent of its forces on a joint basis.[65] Heavy involvement by the army and air force in joint operations is anticipated as well.[66]

The British support joint operations not simply through NATO's integrated command, but by means of bilateral agreements with other Alliance members as well. The British, according to Kolosov, have pursued cooperation with the armed forces of individual West European allies. "Such collaboration is usually carried out on the basis of bilateral agreements that provide for the conduct of a whole series of joint maneuvers." In addition, "in the past decade agreements have been signed and implemented which deal with expanded cooperation among certain branches and contingents of armed forces" (for example, Anglo-Dutch and Franco-British agreements in the naval area). Such agreements are "an important link in England's military partnership with its West European allies."[67]

Also relating to joint operations, it is clear from their writings that Soviet analysts consider the provision of British bases to NATO operations to be critical to NATO's coalition strategy. According to Colonel Artyomov, NATO's leadership believes that Great Britain will be the principal base for supplying and supporting combat actions in the European TVDs: "In wartime approximately 40 percent of the men and equipment [needed] to reinforce the NATO joint armed forces in Europe will be concentrated on British territory."[68] Colonel V. Leskov underscored the significance of the British military infrastructure to the Alliance in similar terms: "Great Britain is located at the intersection of important maritime and air ways and occupies an advantageous military geographic position in Western Europe. The NATO command, taking these particular features into account, has assigned it to a special zone within the system of the European theaters of military action."[69] Colonel Leskov's article includes a map of the key bases and military infrastructure as well, which appear to suggest their priority as targets for Soviet operations against Britain in time of war.

British bases and military infrastructure make a number of contributions to Alliance capabilities. First, they have been critical for providing operational depth for Alliance military forces, especially following France's withdrawal from the integrated military command in 1966. Second, the British bases provide an important staging area for forward-deployed U.S. forces. As Colonel Leskov comments, "According to foreign press reports, the contingent of U.S. troops in Great Britain numbers more than 20,000 people, stationed on more than 20 bases. Given the importance of Great Britain's strategic location, the Pentagon maintains a ramified network of bases, nuclear and conventional weapons warehouses, various staffs, communications hubs, reconnaissance centers, and other installations here."[70]

Naval and air facilities are critical for the resupply of the central front. According to Captain Ivanov, Western strategy assumes that "success in combat operations in the European theater of war will depend primarily on the uninterrupted transfer of troops and military equipment to Western Europe." Ivanov went on to note that Great Britain's naval ports are a critical component in the fulfillment of such plans. Great Britain possesses a "developed network of commercial ports whose further improvement, in order to increase their capacity, meets not only the interests of the nation's economic development, but also the NATO bloc command's plans."[71]

Given the significance the Soviets ascribe to Great Britain's bases and military infrastructure for NATO, it is not surprising that they pay considerable attention to the defense of these facilities as well. For example, Soviet analysts have focused on the plans and equipment Great Britain has provided for air defense of these facilities. It has been noted that the heavy reliance on stationary radars in British air defense means that most of the defenses can be taken out relatively easily in the initial periods of a European war.[72] Similarly, Lieutenant Colonel V. Tamanskii noted that "the existence of imperfect networks of radar stations and vulnerable lines of communication make Great Britain's air defense system ineffective and incapable of accomplishing its tasks in a real combat situation."[73]

In short, Great Britain's military strategy relies heavily on joint operations and allied commitments. The British thus pursue a coalition military strategy to meet national interests. The purpose of this strategy from a British point of view is, of course, the defense of national territory. Kolosov comments that "it is natural that London would view the concentration of its attention on 'defense' of its national territory and its surrounding seas to be the most important priority."[74]

Key Elements of British Security Policy

Soviet analysts focus on three key elements of British security policy. First, the American-British relationship is perceived to be of declining significance to the United Kingdom but still central to British conceptions of security. The main cause of this decreased significance is Great Britain's growing attachment to Europe. Second, and related to the first, Great Britain is playing an increasingly significant role in Europeanizing the Alliance. Great Britain considers the vitality of the NATO coalition to be critical to British security. Great Britain plays an important—perhaps the most important—role of any European power in NATO as an organization and uses NATO to augment its influence within Europe. Armaments cooperation is seen by the British as a critical medium within which to forge a Europeanized Alliance. Third, Great Britain is perceived to be entangled in a significant political struggle over security policy. If Great Britain becomes the first ex-nuclear power as a result of this political struggle, a significant change in the course of British security policy will have been effected. Although the British-Soviet relationship has been a rocky one, a significant change in British security policy might provide new opportunities for the exercise of Soviet influence.

The British-American Relationship

Its special relationship with the United States had gradually become less significant to the United Kingdom. A major factor in this transformation has been Great Britain's membership in the European Economic Community (EEC) since 1973. By joining the EEC, Great Britain increased its trade and foreign economic ties with Europe at the expense of the United States. The British have gradually adopted common European positions on a number of trade issues—positions different from those of the United States. Although a distinctive European identity has animated British foreign economic policy, this identity has affected Great Britain's security policy only partially, a situation that has allowed the United States to retain strong security ties with Great Britain.

Associated with this development of greater European identity has been a growing recognition by the British that their own security policy interests are not identical with those of the United States. For conservative political forces, American-British differences in the Middle East and the conflict over the pipeline deal with the USSR increased their awareness of these conflicting interests. For example, with reference to the 1973 Arab-Israeli war, the Soviet analyst G. A. Vorontsov noted that "the long-term interests of British ruling circles coincided to a greater extent with the aspirations of the West European nations than with U.S. objectives."[75] For the more radical political forces, the conflict over the deployment of cruise missiles in the United Kingdom mobilized public opinion in an anti-American direction. Both conservative and radical political forces were reminded of the differing interests of Great Britain and the United States by the American raid on Libya using F-111s stationed in Great Britain. In fact, were Prime Minister Thatcher not such a committed Atlanticist, it is highly unlikely that permission would have been granted. According to one Soviet report, within her own cabinet Thatcher lacked majority support, and two-thirds of the public did not support the decision as well.[76] In sum, according to Soviet analysts, Great Britain was treated as a tool of American imperialism without regard to its own interests.

A further element in the declining significance of the U.S.-British special relationship has been the decreased attention the United States has paid to Great Britain. As Great Britain has become more European, it has become a less distinctive ally of the United States and must compete with the FRG and

France for U.S. attention. As Davydov et al. commented, "In today's world, England's place within U.S. global foreign policy calculations has been taken by Western Europe as a whole."[77]

Although Great Britain's special relationship with the United States has become less important, it still remains a central component of British security policy in several ways. Above all, the two countries retain a close relationship in the nuclear sphere. The creation of joint nuclear agreements has facilitated the purchase of U.S. nuclear systems, a necessary arrangement for the British since they opted out of independent missile development.[78] In addition, the British engage in joint targeting of their nuclear forces with the Americans. Some Soviet analysts assert American dominance; others assert independence in actual use. According to Vorontsov, "England depends not only on being supplied with American warheads and delivery vehicles, but also on the operational plans of the U.S. command. The latter, in particular, participates in the selection of targets for British strategic weapons."[79]

The special relationship in nuclear matters provides the British with both advantages and disadvantages. According to Trukhanovskii, the advantages include receiving "secret information on nuclear matters, special equipment, nuclear weapons systems, etc. from America." On the other hand, he argues, "this cooperation leaves England seriously dependent on the United States. England's focus on the U.S.-produced Trident-2 missile system is a convincing example of this dependency. Both now and in the future the British command will not be able to determine the precise location of its submarines at sea without U.S. assistance and in order to aim their missiles at a target, data from U.S. satellites will be needed, etc."[80]

Soviet analysts underscore that U.S.-UK nuclear-sharing arrangements provide the British with the opportunity to be the best informed of all America's European allies concerning U.S. nuclear policy. As Kolosov has noted, "The system of agreements has made it possible for British specialists to be up on the development of nuclear armaments in the U.S. and to be better informed in this area than the other NATO West European allies."[81] Also, the British have been put in the position of being able "to influence to some extent the United States's course in elaborating NATO's nuclear strategy, to receive information first-hand about a broad range of problems, and to participate in solving them."[82] From the Soviet point of view, the cost to the British of being well informed is that Great Britain is generally the European power most supportive of American nuclear policy. By being kept in the nuclear ballgame with American assistance, the British are generally the most pronuclear of America's allies within the NATO military organization. As Nikolaev has commented, "Great Britain, being the only West European state belonging to NATO's military organization which possesses and produces nuclear weapons, has always supported all the plans of the U.S. military-political leadership with respect to the further buildup of [NATO's] nuclear might."[83] The British are especially strong supporters for maintaining the U.S. nuclear guarantee to Europe. Kolosov adds that "in supporting close relations with the U.S. in the nuclear weapons sphere, British leaders have attempted to consolidate the 'efficacy' of American 'nuclear guarantees' to England and the other West European NATO allies."[84]

By remaining in the nuclear game, Great Britain not only supports U.S. nuclear policy but also rejects Soviet options for reducing tensions associated with the nuclear competition. As Golubev has noted, "The refusal to support Soviet proposals to freeze nuclear arms and not to use nuclear weapons first is explained to a considerable extent by the fact that the Conservatives set about implementing the program to modernize their 'independent nuclear force,' expecting to increase its yield several times over and to thereby expand NATO's potential for a first strike."[85]

Another area of close cooperation is seen in the British willingness to provide U.S. forces with bases on British soil. Great Britain has allowed the stationing of U.S. nuclear weapons on British soil in the form of both F-111s and cruise missiles. This action is seen by the British as a concrete commitment on their part to bolster the U.S. nuclear guarantee to Europe. They provide bases for significant deployments of U.S. conventional forces as well.

The various ties between Great Britain and the United States have meant that Great Britain is America's staunchest military ally in Europe. The British have used this position to try to ensure that the United States maintains a policy conducive to British interests as articulated in the positions of the British government. This is especially clear in the nuclear area, where the British have sought to maintain the American nuclear guarantee to Europe. For example, while the Americans at Reykjavik seemed to support a serious reduction of nuclear forces, the British sought after Reykjavik to encourage the United States to return to previous positions.

In fact, one Soviet commentator chose to contrast the position taken by President Reagan at Reykjavik with that taken by Prime Minister Thatcher during her visit to Moscow in March 1987. Gennadii Gerasimov noted that "President R. Reagan believes that the concept of nuclear deterrence is immoral and—even if only in words which differ from the Pentagon's practice—dreams of living to see a world without nuclear weapons. Britain's Prime Minister M. Thatcher holds a different opinion." She believes in nucleophilia, which "is an invitation to continue the arms race."[86]

Although Thatcher's support for the final INF agreement has been duly noted, the Soviet press has also underscored her uneasiness about a 50 percent reduction in strategic arms.[87] Moreover, a TASS report of Geoffrey Howe's meeting with Gorbachev in Moscow in February 1988 charged that Great Britain was leading the effort to prevent progress in disarmament:

"While generally approving the INF Treaty, the leaders of NATO countries have been acting as though the situation in Europe since its signature has only worsened."[88] Proof of this reaction, TASS claims, can be seen in joint military arms production efforts, defense of the deterrence doctrine, and in plans to compensate for the weapons systems removed under the INF agreement. With respect to this latter issue, the Soviets see Great Britain in the forefront of these plans, for which the country is strongly criticized. *Pravda* reporter Aleksandr Lyutyi has, in fact, quoted a *New Statesman* article that indicates that NATO plans to increase the number of U.S. nuclear warheads in Great Britain from 775 to 1,193.[89] Great Britain is thus depicted as the United States' most willing ally in nuclear issues.

In short, the special relationship works both ways. On the one hand, the Americans use the relationship to try to make the British spearhead their efforts to influence Western Europe. On the other hand, the British seek to influence the Americans in ways favorable to their policy and increasingly rely on their European connection to try to influence the Americans more effectively. Although most Soviet analysts focus on the first dimension of the special relationship, increasing attention is being paid to the latter as well.

Great Britain and the Europeanization of the Alliance

Soviet analysts argue that the British have a twofold approach to the Europeanization of the Alliance. On the one hand, the British draw on their position in NATO to try to act as the interlocutor between Europe and the United States. In fact, during the 1988 U.S. presidential campaign, one Soviet commentator suggested that Great Britain was seeking to lead NATO while the United States was occupied with its political transition. More generally, the British try to use their position in NATO to generate greater influence within Europe. As Kolosov commented, "Striving in every way possible to increase England's influence with respect to its European Community partners, British leaders have frequently counted on the fact that further fulfilling its role as the leading West European military power in NATO will allow Britain to have the necessary influence in other areas too."[90] On the other hand, the British seek to encourage the development of a "European pillar" within NATO. To facilitate this process, they helped create the Eurogroup and the Independent European Program Group (IEPG) and have developed extensive ties with other European states in joint armaments production.

In the Soviet view, the British play the preeminent European role in NATO, having the greatest degree of representation in the command structure. As Kolosov has noted, "in comparison to the Americans' role in the military organization of the North Atlantic Alliance, the position of the British generals and admirals can appear relatively modest. But actually their influence in

NATO, while of course lagging behind the Americans', noticeably surpasses the influence of the other allies. This is manifested primarily in the allocation of the higher posts in NATO's integrated command."[91] They have also played a central role in the various NATO committees. Kolosov has argued that all this high-level influence, combined with "the significant 'presence' of lower rank British officers and specialists in NATO's headquarters, commands, and various committees, is called upon to protect the specific interests of England's government and military leadership and not to permit a diminution of its influence within the bloc's military organization."[92]

Based on their position in NATO, the British want to act as Europe's "special voice" in dealings with the United States. Davydov et al. commented that

> the long-term objective of U.S. strategy is to establish new relations with a unifying Western Europe on the model of the former "special relationship" with England, and this stipulates the need for and importance to Washington of continuing close Anglo-American relations. Such calculations by American ruling circles are reinforced by London's mutual aspiration to exploit close Anglo-American relations in order to gain a strong position within a unifying Western Europe, primarily with its main EEC partners, France and the FRG. Both the U.S. and London appraise the remnants of the "special relationship" as important assets which can bring significant dividends to both partners in the event of the further realization of the ideas of the military and political integration of Western Europe.[93]

Illustrative of Great Britain's trying to act as a conduit for Europe in dealing with the Americans has been its conduct in East-West relations. Shchelokova commented that "British diplomacy is striving to emerge as the coordinator of the [EEC's] foreign policy actions in East-West relations in order to strengthen the positions hostile to the socialist countries, as well as to smooth over differences of opinion on these issues between Western Europe and the United States."[94] But the new Soviet leadership also perceives possibilities for using this role to its own advantage; as discussed earlier, the Soviets played to this role during Gorbachev's London visit in December 1987.

The British have also sought to draw on their position in NATO in order to expand their influence within Europe. Although Great Britain's relative position has declined economically within Europe during the postwar period, it has remained an important military power. According to Soviet analysts, the British consistently spend more on defense than any other European power. They do this not only to support "fading dreams of imperialist glory," but to maintain a key foundation for the exercise of influence in Europe as well. As Belyaev underscored, "The ambitious plans of the Conservatives, dreaming about the former greatness of the British empire and aspiring to play a leading role in the Western world, play no small role in the build up of Great Britain's military might."[95]

As the European power center has become more assertive, the British have played an important role within that power center. According to Madzoevskii and Khesin, "England occupies quite a definite place within the U.S.-West European-Japanese triangle and acts, within this framework, as an integral component of the West European center."[96] These authors explain further that the British have sought to Europeanize the Western Alliance through the formation of such institutions as the Eurogroup: "A definite 'Europeanization' of its [Britain's] nonnuclear component is taking place under the auspices of the Eurogroup's programming, which is uniting the majority of the European members of NATO (without the participation of the U.S. and Canada)."[97] The NATO Eurogroup is perceived by Kolosov as playing an "extremely vital" role in coordinating collaborative efforts among the West Europeans and has "contributed to the increase of their contribution to the build-up of NATO's military might and to the development among them of military-policy, military, and military-industrial ties."[98]

The Eurogroup, however, has acted only in certain spheres to promote collaboration. Kolosov has detailed that

> attention has been concentrated on tasks, the execution of which could be of "local" importance. The more general and substantial problems of coordinating the national military planning of the participating nations and coordinating the programs to develop the armed forces and military training have, to all intents and purposes, remained outside the sphere of the Eurogroup's activities. Of course, the constant efforts of the Eurogroup members to expand coordination in military training and material-technical support have established the basis for producing new tasks that are more military-policy than military-technical tasks. [99]

Thus, the Eurogroup has focused its attention primarily on coordinating efforts in what Kolosov calls military-technical rather than military-policy tasks.

The formation of the Eurogroup reflects the basic British approach to Europeanization. As Madzoevskii commented:

> As is known, London is the founder and leader of NATO's Eurogroup. Naturally, England's initiative in establishing this organization directly reflected the "Eurocentrist" tendencies in its own policy. But at the same time, from the viewpoint of British and American ruling circles, an important function of the Eurogroup lies in adapting NATO to the changing correlation of forces between Western Europe and the United States and to thereby prolong the life of the North Atlantic Alliance.[100]

The linkage among the Eurogroup, Europeanization, and the significance of European—including British—conventional forces to greater inter-European cooperation has been clearly articulated by Kolosov.

In the military arena, England, the FRG, Italy and the other West European members of NATO provide the main portion of general purpose forces that are focused on executing the bloc's tasks.... There is a growing potential for developing more durable and long-term ties among the national ministries of defense and the armed forces in the most diverse fields—from their material-technical support to the conduct of joint maneuvers, and the elaboration of, to all intents and purposes, unified operational tactical plans for conducting a war.

To this should be added that a significant portion of the NATO infrastructure in Western Europe... is under the formal authority of the West European nations.... The potential for coordinating the West European allies' contributions to the NATO program is not limited to the development of the infrastructure. As the experience of the Eurogroup's activities have shown, such coordination can encompass coordination of issues connected with rearmament, armed forces' cooperation, combat training, and material-technical support.[101]

In other words, the British have not been receptive to Europeanization approaches outside the NATO context. According to Kolosov, "England, the FRG, and Italy have very unequivocally underscored the importance of focusing West European cooperation on strengthening NATO and the prematureness or even undesirability of assigning it [cooperation] a large amount of independence except, possibly, in the military-industrial field."[102]

From the Soviet point of view, the reason Great Britain has rejected Europeanization outside a NATO context has been the priority the British place on NATO's coalition military strategy.[103] By and large the British perceive their forces to be most effective when operating jointly with NATO's forces. The British have drawn the Americans into a tighter nuclear guarantee by allowing deployment of U.S. nuclear weapons on British soil and by promoting nuclear interdependence with the United States. The British have, however, retained the ultimate capability to use their nuclear weapons to serve national interests. Moreover, an article written by a noted Soviet analyst of West European integration, G. Kolosov, indicates that the ultimate goal of West European cooperation is the formation of a Euro-deterrent force. Nevertheless, he continues, because the issue is so complicated, for the foreseeable future Western Europe will have to continue to rely on the U.S. nuclear guarantee.[104]

The most concrete manifestation of the Europeanization of British security policy has been in the armaments sphere. According to Soviet analysts, there has been a dramatic shift away from national or British-American cooperation in armaments production to European cooperation.[105] Soviet analysts have focused on a number of key European coproduction efforts in which Great Britain has been involved, most notably in the aerospace field.[106] The Jaguar aircraft was the result of successful Anglo-French cooperation. The Tornado has been built by a consortium of West German, Italian, and British firms. The EFA will also be built by a similar consortium of European firms within which British aerospace firms will play a prominent role. Soviet analysts note that such projects are indicative of West European, including British, efforts to reduce their dependence on the United States for military equipment.[107]

Cooperation in the armaments field is not an easy process. There has been clear conflict among British firms regarding the desirability of such cooperation. For example, in the aerospace field Kolosov has noted that

> the leading British firms... were in favor of an ever greater focus on implementing joint arms production projects. On the other hand, the small firms in this field, greatly dependent upon military orders, ... evaluated the prospects for their own collaboration fairly pessimistically and emphasized the advantages of national programs.... Thus, one can hardly say that the policy of England's active participation in joint West European arms production was unconditionally supported by all British defense industries.[108]

The British have relied on arms sales as a major means of supporting their military production system.[109] Increasingly, European-coproduced weapons play a key role in British arms exports—for example, the sale of the Tornado aircraft to Saudi Arabia. Coproduction enables British firms to build more advanced weapons than would otherwise be possible, which gives them the ability to compete in the global arms market.

To facilitate the contribution of armaments coproduction to Europeanization, the British encouraged the Eurogroup to create the IEPG. This organization, which includes France among its members, was created in 1976 to provide an institutional framework for the discussion of armaments cooperation. The focus of the IEPG has been limited primarily to the preparation of recommendations by specialists for arms cooperation and their acceptance by representatives from the participating nations. The involvement of France in the IEPG's activities represents an important step forward. Kolosov has argued that "even if the IEPG's activities are limited to coordinating rearmament planning and searching for possible joint projects... in this case, too, one could speak of an important qualitative improvement in the development of mechanisms to regulate West European military-industrial collaboration—the formation of a center with the participation of all three leading states [Great Britain, France, and West Germany]."[110]

Despite growth in intra-European cooperation over the years, formidable barriers to a fully cooperative role by Great Britain remain. The British government is susceptible to pressures from the military-industrial complex to "buy British." A clear example of this phenomenon was the heavy investment in the development of the aborted Nimrod early-warning system. In addition, the desire of each key West European state for a fully developed military-industrial system naturally leads to conflict. Especially notable in this regard has been Anglo-French competition within the EFA project, which has led to France's elimination from the project.

The Soviets expect the British orientation toward Europeanization to emphasize the nuclear component of European defense. Many Soviet analysts of post-Reykjavik developments were struck by the extent to which the European conservative governments sought to work together to ensure that the Americans stayed on the nuclear path. Typical of this concern was the report by an *Izvestiya* correspondent concerning Thatcher's visit to Paris in November 1988 to enhance her ability to influence the Americans on nuclear disarmament issues.

> The talks dealt primarily with the results of the Soviet-American meeting in Reykjavik. Paris and London are expressing their "concern" over the prospects that have arisen for eliminating nuclear weapons and, in particular, medium-range missiles in Western Europe. In unison, the two capitals are advancing the theory that since the socialist countries supposedly enjoy an advantage in conventional weapons, then nuclear disarmament, they say, is fraught with danger for Western Europe.... It is being noted here that before leaving for Paris, Margaret Thatcher made a voyage across the ocean, where she shared her "apprehensions" on behalf of the Western Europeans.[111]

One other issue that incorporates the nuclear dimension and relates to the post-INF environment is the expansion of Anglo-French defense cooperation. A plan to create a Paris-London nuclear axis was mentioned in a brief Soviet report on a meeting between Prime Ministers Thatcher and Jacques Chirac shortly after the former's return from Moscow in April 1987.[112] The month the INF agreement was signed saw publication of an article by A. Lyutyi on Anglo-French efforts to compensate for the withdrawal of the U.S. INF systems. Specifically, he reports a significant expansion of their bilateral cooperation, including the development of an air-launched cruise missile.[113]

The British Security Debate

In assessing the British security debate, Soviet analysts focus the bulk of their attention on developments within the Labour party, and only secondarily on the Social Democratic Party (SDP), Liberal Alliance and the Conservative party.[114] The reason for this emphasis is not difficult to understand. The Soviets believe that the greatest probability of developments favorable to their interests would occur under a future Labour government and much less so with the Alliance or the Conservatives.

In covering the June 1987 election campaign in Great Britain, Soviet commentators focused on several key themes. Public opinion polls showed that Prime Minister Thatcher and her Conservative party enjoyed a strong advantage over the opposition parties when the election date was announced, primarily because of Great Britain's relatively sound economic performance in the months preceding this announcement and Tory successes in local elections during this same period. As Soviet (and other) analysts indicated, however, Thatcher began to lose this advantage midway through the campaign, finding that the public was not necessarily pleased with her domestic policies and, specifically, did not support many of her social and economic programs. The Labour party in particular proved quite effective in criticizing her performance on domestic issues. Thatcher's response was to attack Labour's defense policies sharply, arguing that unilateral nuclear disarmament would leave Great Britain unprotected in the face of the Soviet threat.

The strategy of shifting some of the campaign's focus to defense issues helped Thatcher regain lost ground in the public opinion polls. Interestingly, although Soviet commentators criticized the prime minister's anti-Soviet rhetoric, they did not directly support Labour's defense policies. Soviet newspaper articles and television broadcasts at this time remained neutral when describing Labour's antinuclear positions. In fact, Nikolai Gorshkov refused to comment directly on this issue, even when specifically asked how the Soviet Union viewed Labour's defense platform. Gorshkov stated that it was too sensitive a topic and that he would not wish to be perceived as interfering in the election campaign. The strongest statement he was willing to make was that "we have to accept new thinking according to which less arms means more security; and from this point of view Labour's defence programme appears to show the signs of this new thinking."[115] This reluctance on the part of the Soviets to "interfere" can certainly be attributed in part to their desire to point an accusatory finger at President Reagan, who publicly stated that Labour's defense policy was a serious mistake.

In an assessment of the election results, *Pravda*'s leading analyst of Great Britain, Arkadii Maslennikov, sets forth four reasons for the Conservative victory. First, Thatcher was able to convince the British public that they were financially better off than when she first took office and that continued future improvements depend on the Tories staying in power. Second, the media coverage of the campaign generally favored the Conservative party. Third, Thatcher played up British patriotism and anti-Sovietism and stressed the fact that Labour's defense policies would lead to an unprotected Britain. Finally, Maslennikov singled out the problem of fragmentation and schisms within the opposition parties as a significant factor detracting from their ability to defeat the Conservatives.[116]

The decreasing consensus on nuclear policy has led to increased debate within Great Britain on security policy. The main stimulus to the debate was the NATO decision to deploy new American cruise missiles in Great Britain. As V. Kelin commented:

> The growth in the West, including in the British Isles, of alarm [due to U.S. deployment of cruise missiles and Pershing IIs] among broad segments of the masses, in various political circles, about their present and future has introduced a definite, new factor into the situation. Suddenly calls have begun to rain down from official circles in the Western states to begin a dialogue between the Soviet Union and the United States, between East and West.[117]

This deployment in Great Britain led to a revitalization of the British peace movement. Pressures from the peace movement induced the Labour party to become more radical in its own views on nuclear disarmament. An additional stimulus to the debate has been the British decision to buy Trident submarines to replace the current Polaris force. The Soviets consider Trident to be an unwarranted expansion of the British strategic force that alters the fundamental character of the British deterrent. Soviet analysts have noted that the Labour, Liberal, and SDP parties have all rejected Trident.[118]

As already noted, developments within the Labour party, especially its program for unilateral disarmament, interest Soviet analysts the most. Labour is committed to canceling Trident, removing U.S. cruise missiles, and eliminating all U.S. nuclear weapons from British territory. Soviet analysts perceive that the Labour party is serious about its program for such unilateral disarmament. The Labour leadership has become more radical, and pressure from the mass peace movements to maintain a radical position has also played a significant role. Even Neil Kinnock's attempt to temper the party's defense platform met with a flat rejection. Nonetheless, some doubts persist about whether Labour would actually implement its program if it came to power. Ryzhikov has argued that the Labour party in opposition "has taken a positive position on many issues that are important to the British people. This does not, however, by any means signify that all the promises made by the Labourites will automatically be implemented should they come to power. Unfortunately, the experience of history has taught that the words and the deeds of the Labourites are [two] different things."[119]

In spite of Labour's strong antinuclear sentiment, the party still seems committed to adherence to the Western Alliance. When in power in the 1970s, the Labour party proved to be stronger Atlanticists than even the Conservatives. But the Labour party of the 1970s no longer exists. A Labour party government of the 1980s could be expected to remain within the Alliance but to challenge U.S. positions on nuclear and other issues within the Alliance.

The center coalition of the Liberals and SDP was deeply divided on security issues, a fact that contributed to the breakup of the coalition. While Soviet analysts treated Liberal and Labour positions as virtually identical, the SDP was perceived (prior to its merger with the Liberal Party) to be a different animal altogether. The SDP was seen as anti-Trident, pronuclear, pro-European, and only mildly anti-American. If the SDP had become part of a government, the Soviets would have expected to see an effort to spur Europeanist as opposed to Atlanticist tendencies. Following the coalition's demise, Soviet press accounts have paid virtually no attention to these or any other parties except Labour and the Conservatives.

The Soviet view of the Conservative party has been heavily dependent on who the party leader is. Under Edward Heath in the 1970s, the Tories were seen to have strong Europeanist leanings. Vorontsov argued that under Heath, "the Conservatives leaned strongly toward a European orientation . . . which was dictated by England drawing nearer to the EEC and by the presence of sharp contradictions with the United States."[120] In contrast, under Thatcher in the 1980s, Atlanticism has predominated. But even under Thatcher there remain strong undercurrents of Europeanist tendencies in the party, as evidenced by the Westland affair.[121] Michael Heseltine, then Minister of Defense, was perceived to be a strong Europeanist and, as such, in opposition to Thatcher's Atlanticism. Despite the conflict between Atlanticism and Europeanism within the Tory party, both elements are strongly committed to Great Britain's remaining a nuclear power. In fact, only the Tory party supports the Trident program.

From the Soviet point of view, continued adherence to Great Britain's nuclear stance impedes Anglo-Soviet relations. The Thatcher government insists on pursuing its nuclear modernization program and excluding its significant increase in nuclear warheads from an arms control agreement. Inevitably, the stalemate between the Soviet Union and the United Kingdom on nuclear issues impedes positive relations between them.

Soviet analyses of the internal British security debate are conducted with an eye to the exercise of Soviet influence. As Kolosov has carefully noted:

> One should emphasize that there does exist a quite real potential to influence the development and conduct of England's military-political course in the context of its greater conformity to the genuine interests of ensuring the nation's security. Here, a great deal depends on who is in power. The Labour leaders, with all their vacillations and inconsistencies during the 1970s, displayed a greater readiness to act in this direction than did the Conservative leaders. It would, of course, be an oversimplification to assume that within the leadership of the Conservative party there are no politicians who are inclined to take the step, sooner or later, to make suitable modifications in at least individual components of the British military-political course, but all the same, such changes are still more probable in the event that Labourites come to power, possibly in alliance with the other centrist parties.[122]

Clearly the Soviets hope to encourage or reinforce Labour's antinuclear policy. In 1985 Lieutenant Colonel Roshchupkin outlined how the Soviets might do this:

> The Soviet Union would be prepared to reduce and physically liquidate the same portion of its medium-range missiles in the European part of the USSR as would correspond to the number of nuclear missiles liquidated by the British side. Britain's implementation of complete nuclear disarmament, with the abolition of corresponding foreign bases, would create the conditions whereby the USSR would guarantee that its nuclear weapons would not be targeted against British territory. In the event of an official decision by Great Britain on nuclear disarmament, the entire series of questions arising as a result of this move that are related to Soviet-British relations in the military sphere could become the subject of discussions and a corresponding agreement between the USSR and Great Britain.[123]

Gorbachev made this proposal officially in May 1986 to a group of British parliamentarians, including Denis Healey, the Labour "shadow foreign minister. The two key elements of the proposal were the following: If Great Britain would liquidate its nuclear weapons, the Soviets would cut their weapons by an equivalent amount. If Great Britain removed U.S. nuclear weapons from its territory, the former would not be targeted by Soviet nuclear weapons.[124]

The American raid on Libya in 1986 heightened British concern about U.S. bases in Great Britain. More generally, this issue is one that could divide the British from the Americans and therefore be useful to the Soviets. As Trukhanovskii noted, "British public opinion has for many years been worrying about the problem of who [will decide] and how the decision will be made to use U.S. bases, in the event of war. There exists the fear—and a very well-founded one—that the Americans will resolve this matter on their own. This means that England could be drawn into a nuclear war not by its own government, but by the U.S. military. And this threatens the extinction of the country."[125]

If Europeanist tendencies are accelerated in post-Thatcher Great Britain, the potential for more favorable Anglo-Soviet relations is enhanced. To the extent that European states shift away from Atlanticism, their independence is increased and the potential for establishing mutually beneficial relations with the Soviet Union are augmented. The British, however, could pose a dangerous wild card in the acceleration of Europeanist tendencies. Because the British have relied heavily on their military power to augment their influence in Europe and have sought to create a European pillar within and not outside the Alliance, the British might well seek to create a militaristic Europe antithetical to Soviet interests.

Even under an Atlanticist leader like Thatcher, American and British interests are not identical, which provides an opportunity for the exercise of Soviet influence to support mutually beneficial relations. First, Great Britain did not support the U.S. effort to abrogate the gas pipeline deal between Europe and the Soviet Union. According to Golubev, Great Britain resisted the American attempt to apply sanctions against British companies supplying equipment to the Soviet Union because "both the prestige and vital economic interests of Great Britain were affected."[126] Second, even the Conservatives could not jettison the effort on the part of the West to try to reach arms control agreements with the Soviet Union. Golubev commented that "the Conservatives could not openly declare their rejection of the policy of disarmament. They felt strong internal political pressure."[127] He added elsewhere that "the experience of the policy of detente, which meant much more for Great Britain as a European nation than it did for the United States, and the need to continue dialogue on limiting the arms race—whose rejection could have serious internal political consequences for the Conservatives—limited the possibilities for the practical manifestation of anti-Sovietism."[128]

Clearly the way the Soviets played the British and French nuclear weapons issue in the months preceding the INF agreement reflected a desire to gain influence for having made a concession by not insisting on the inclusion of these systems in the INF agreement. As one Soviet commentator noted, "Discussing medium-range missiles in Reykjavik, the Soviet Union agreed to leave aside the British and French nuclear forces, and it was a large concession on the part of the USSR."[129] Another Soviet commentator reminded his Soviet listeners that "this does not at all mean that the question of these forces is removed from the agenda altogether; for they exist and represent a real quantity that must be taken into account."[130]

Soviet analysts noted the response of the various parties to the Soviet concession. The Conservatives rejected the concession because of their attachment to nucleophilia.

> It is absolutely clear that the aim is to defend Britain's nuclear potential and to preserve its membership in the privileged "nuclear club" in order albeit somehow to be different from other Western states. After all, London had at one time been promising "with time to join" the USSR and the USA if they agreed to a substantial reduction in their nuclear forces. Now that real opportunities are appearing for beginning the process of nuclear disarmament, and a real threat has arisen to Britain's imaginary privileges stemming from nuclear status, the promise once made has become quite undesirable and a burden for the Tory government.[131]

Opponents of the Tory policy are noted by the Soviets as providing much more flexibility on the issue of the British forces. For example, Neil Kinnock has argued, "it is incomprehensible how Mrs. Thatcher intends to achieve an expansion of mutual understanding and co-operation and at the same time is threatening the Soviet Union with nuclear weapons."[132]

Conclusions

The Soviet assessment of British security policy is based on the conviction that it forms a critical component of overall Western security policy and that, far from being an isolated phenomenon, it serves as a reflection of broader developments. As such, it provides a channel for influencing general Western security trends.

The real significance of Great Britain in the Soviet Union's East-West policy has varied. At times, developments in Great Britain have been watched with special interest; at other times, the interest has been less marked. In general, however, Soviet analysts have been especially interested in Great Britain's role as a special ally of the United States in the Western Alliance, since it is from this vantage point that the British still have an opportunity to influence U.S. policy and thereby play a greater role than they could play alone.

The issue of nuclear deterrence in Europe has also been a matter of special concern, and Great Britain's role in maintaining the European deterrent system has naturally aroused keen interest. Great Britain's possession of an independent nuclear force has placed the British "objectively" on the side of the Americans in maintaining the nuclear basis of deterrence. But it has also imposed a constraint on U.S. policy by encouraging the Americans to think in nuclear terms in their reaction to European security issues more than might otherwise have been the case. From the Soviet point of view, the attitudes of the British government—reinforced by those of the French—will seriously limit the efficacy of continued nuclear disarmament efforts.

In order to break the deadlock on the Western side over the issue of nuclear deterrence and its future, the Soviets will seek to influence the British debate as much as possible. But under Gorbachev, they are also seeking to change Western attitudes toward the Soviet Union more generally. By gaining Western acceptance of the Soviet Union's domestic modernization program as a positive development and by highlighting the peaceful implications of the program for the West, the Soviets also hope to diminish Western threat perceptions. By maintaining and extending their public diplomacy, the potential of which has been underlined by the series of arms control proposals submitted since early 1986, they will attempt to ensure that the reduction of nuclear weapons in Europe is not matched by any significant upgrading of the conventional deterrence system.

Great Britain itself is considered to play an essential role in the conventional defense efforts of the Western Alliance; this is due not only to the strength of Great Britain's own forces, but even more so to the military infrastructure and network of bases the British have placed at the disposal of the Alliance through their U.S. connections. As noted earlier, the Soviets will seek to exploit any tensions between the United States and Great Britain over issues related to these bases.

But the Soviets have also noticed important deficiencies in Great Britain's conventional forces. Most notably, much of the country's present equipment is known to be aging, and the new weapons that are designed to replace it are often too costly to be deployed rapidly or in large numbers. The recurrent tension between the nuclear and conventional budgets has also been carefully noted by Soviet analysts. From this perspective, the financial problems created by Great Britain's nuclear modernization program are perceived to be working to the Soviet Union's advantage, because they will almost certainly mean a corresponding reduction in Great Britain's conventional force. The Soviet Union will seek to influence attitudes on the British left in order to ensure that, if the left comes to power, any savings in government spending that result from the elimination of nuclear weapons will not be diverted into conventional modernization.

Great Britain is also considered to have a pivotal function in the emergence of a new U.S.–West European relationship, in which the latter will play a greater political and military role. Here again, it is the broader relationship between the Western states that provides the basis for evaluating the significance of British actions. By stimulating a militaristic approach to the Europeanization of the Alliance, the British could prove especially troublesome to the Soviets. They might encourage their fellow Europeans to develop their own nuclear forces and to cooperate more effectively in joint military operations within the NATO framework. In any event, the kind of Europeanization that Great Britain appears to favor is antithetical to Soviet interests—or, as Soviet analysts put it, antithetical to the "genuine" interests of the British people, which would be better served, they claim, by a system of all-European cooperation.

Finally, the Soviets will be sure to follow closely Great Britain's internal security debate and to encourage, as opportunities permit, a defense policy that is both nonnuclear and more accommodating in conventional terms. If the British could be persuaded to relinquish their nuclear weapons while also refraining from competing over conventional arms, their example might induce broader changes throughout the West.

In this way, the significance of British security policy lies in its impact on broader East-West policies. If the British remain firm in their defense policy, one of the most important barriers against Soviet pressure will remain intact. If they give way, new and better opportunities will have been created for the expansion of Soviet influence, and the Western Alliance itself will have been seriously weakened.

Notes

1. Research for this chapter was completed in December 1988. The authors are indebted to Daniel Peris for his research assistance in providing updates for this material since mid-1987.
2. S. P. Madzoevskii and E. S. Khesin, eds. *Velikobritaniya [Great Britain]* (Moscow: Mysl', 1981), p. 365.
3. Ibid., p. 358
4. G. A. Vorontsov, *SShA i Zapadnaya Evropa: novyi etap otnoshenii [The U.S. and Western Europe: A New Stage in their Relations]* (Moscow: Mezhdunarodnye otnosheniya, 1979), p. 228.
5. Madzoevskii and Khesin, *Velikobritaniya*, p. 357.
6. Ibid., p. 359; Vorontsov, *SShA i Zapadnaya Evropa*, p. 220.
7. S. P. Madzoevskii "'Europeanism' and 'Atlanticism' in London's Policy," in "Great Britain in the Mid-Seventies," *MEMO*, 4 (1975), p. 125.
8. S. P. Peregudov, *Leiboristskaya partiya v sotsial'no-politicheskoi sisteme Velikobritanii [The Labour Party in Great Britain's Socio-Political System]* (Moscow: Nauka, 1975), p. 139.
9. A. Likhotal', "The Labourites and European Security," in "Great Britain in the Mid Seventies," *MEMO*, 4 (1975), p. 126; Madzoevskii and Khesin, *Velikobritaniya*, pp. 364–365; and Peregudov, *Leiboristskaya parityva*, pp. 139–140.
10. T. Nichkova, "'Defense' Policy," in "Great Britain in the Mid-Seventies," *MEMO*, 4 (1975), p. 126.
11. V. A. Ryzhikov, *Britanskii leiborizm segodnya: teoriya i praktika [British Labourism Today: Theory and Practice]* (Moscow: Mysl', 1984), p. 228.
12. V. G. Trukhanovskii and N. K. Kapitonova, *Sovetsko-angliiskie otnosheniya, 1945–1978 [Soviet-British Relations, 1945–1978]* (Moscow: Mezhdunarodnye otnosheniya), pp. 246–247.
13. Ryzhikov, *Britanskii leiborizm*, p. 244.
14. Ibid., p. 258.
15. Madzoevskii and Khesin, *Velikobritaniya*, pp. 359–360.
16. A. V. Golubev, "Soviet-British Relations at the Turn of the Eighties," *Voprosy istorii*, 7 (1984), p. 53. See also G. V. Kolosov, "The Military-Political Aspects of the West European Integration Process," in N. S. Kishilov, ed., *Zapadno-evropeiskaya integratsiya: politicheskie aspekty [West European Integration: Political Aspects]* (Moscow: Nauka, 1985), p. 245.
17. A. Maslennikov, "Hanging onto Uncle Sam's Coattails," *Pravda*, March 19, 1986, p. 5.
18. A. Maslennikov, "What's Behind the Crisis?" *Pravda*, January 13, 1986, p. 5.
19. V. G. Trukhanovskii, "England's Nuclear Policy, 1979–1984," *Voprosy istorii*, 6 (1985), p. 25.
20. Ibid., p. 28.
21. Ibid., p. 46.
22. V. P. Nikhamin et al., *Vneshnyaya politika Sovetskogo Soyuza [The Foreign Policy of the Soviet Union]* (Moscow: Politizdat, 1985), p. 239.
23. Ibid., p. 240.
24. "Results of Mrs. Thatcher's Visit: Further Comment," *SWB*, April 4, 1987, pp. 3–4.
25. V. Bol'shakov, "What Are They Afraid Of?" *Pravda*, January 17, 1987.
26. "Mrs. Thatcher's Television Interview on 31st March," *SWB*, April 2, 1987, pp. 4–13.
27. A. Lyutyi, "Relations Improved," *Pravda*, December 9, 1987.
28. TASS, "Soviet-British Talks," *Pravda*, February 16, 1988.
29. Capt. 1st Rank A. Karemov and Col. (Ret.) G. Semin, "Certain Provisions of the Military Doctrines of the Main NATO European Countries," *ZVO*, 6 (1983), p. 10.
30. Ibid., p. 9.

31 G. V. Kolosov, *Voenno-politicheskii kurs Anglii v Evrope [England's Military-Political Course in Europe]* (Moscow: Nauka, 1984), p. 113.
32 Capt. 2nd Rank Yu. Marov and Capt. 3rd Rank A. Biryusov, "Certain Results of the Falklands Conflict," *ZVO*, 5 (1984), p. 16.
33 Capt. 2nd Rank Yu. Galkin, "The Air Defense of British Expeditionary Forces (during the Anglo-Argentinian Conflict)", *ZVO*, 3 (1983), pp. 64–67.
34 Lt. Col. A. Aleksandrov and Capt. 3rd Rank S. Grechin, "The Falklands: A Recurrence of British Colonialism," *ZVO*, 10 (1982), p. 14.
35 Capt. 2nd Rank Yu. Galkin and Capt. 3rd Rank S. Grechin, "Great Britain's Naval Forces," *ZVO*, 2 (1983), p. 69.
36 Lt. Col. S. Anzherskii, "Great Britain's Armed Forces," *ZVO*, 9 (1984), Part 2, p. 19.
37 Galkin and Grechin, *ZVO*, 2 (1983) p. 75.
38 Capt. 1st Rank (Res.) V. Mar'in, "British Invincible-class Anti-Submarine Cruisers," *ZVO*, 3 (1980); Capt 1st Rank (Ret.) N. Petrov, "Broadsword-class Guided Missile Frigates," *ZVO*, 8 (1983), pp. 72–73.
39 A. I. Utkin, "On Expanding the Spheres of NATO's Actions," *MEMO*, 5 (1987), pp. 34–42.
40 Col. V. Artyomov, "Great Britain's Air Force," *ZVO*, 10 (1982), p. 43.
41 Anzherskii, *ZVO*, 9 (1984), p. 17.
42 Ibid., p. 18.
43 Artyomov, *ZVO*, 10 (1982), p. 43.
44 Col. E. Nikolaenko, "Great Britain's Military Budget for FY1983–84," *ZVO*, 11 (1983), p. 23.
45 Col. P. Shiryaev, "The British Air Force Command in the FRG," *ZVO*, 3 (1985), p. 39.
46 Lt. Col.–Engineer P. Ivanov, "The Anglo-French Fighter, the Jaguar," *ZVO*, 5 (1980), p. 51.
47 Col.-Engineer (Res.) K. Borisov, "The Program to Develop a New Fighter for Great Britain's Air Force," *ZVO*, 5 (1980), pp. 57–58.
48 Lt. Col. S. Anzherskii, "Great Britain's Ground Troops," *ZVO*, 10 (1985), p. 27.
49 Lt. Col. S. Anzherskii, "Great Britain's Armed Forces," *ZVO*, 9 (1980), p. 27.
50 Anzherskii, *ZVO*, 10 (1985), p. 27.
51 Capt. 2nd Rank G. Perov, "Prospects for Developing British SSBNs," *ZVO*, 6 (1981), p. 66.
52 Nikolai Gorshkov, "Labour's Defence Programme Shows 'Signs of New Thinking'," *SWB*, May 13, 1987, p. 6.
53 Ibid.; Capt. 2nd Rank S. Grechin, "Rearming the British SSBNs," *ZVO*, 12 (1985), p. 87.
54 See, for example, Melor Sturua, "Independence or Shortsightedness?" *Izvestiya*, March 19, 1986, p. 4; Lt. Col. V. Roshchupkin, "Ruled Out: The Nuclear Forces of England and France in NATO's Strategy," *KVS*, 5 (1985), p. 84.
55 N. Nikolaev, "Great Britain: Following the Lead of the U.S.' Aggressive Course," *ZVO*, 7 (1984), p. 12.
56 Perov, *ZVO*, 6 (1981), p. 68.
57 Ibid., p. 66.
58 Kolosov, *Voenno-politicheskii kurs*, p. 64.
59 Kolosov, in Kishilov, ed., *Zapadno-evropeiskaya integratsiya*, p. 243.
60 Grechin, *ZVO*, 12 (1985), p. 87.
61 G. V. Kolosov, "Great Britain's Military-Political Course at the Start of the Eighties," *MEMO*, 1 (1983), p. 98.
62 Ibid., p. 99.
63 Sturua, *Izvestiya*, March 19, 1986, p. 4.
64 Kolosov, *Voenno-politicheskii kurs*, p. 27.
65 Galkin and Grechin, *ZVO*, 2 (1983), p. 75.

66 See, for example, Lt. Col. I. Vladimirov, "Great Britain's Military Policy," *ZVO*, 7 (1981), p. 10.
67 Kolosov, *Voenno-politicheskii kurs*, p. 110.
68 Artyomov, *ZVO*, 10 (1982), p. 43.
69 Col. V. Leskov, "Great Britain: Geographical Conditions, State System, Economy, and Elements of Its Infrastructure," *ZVO*, 5 (1980), p. 23.
70 Ibid., p. 26.
71 Capt. 1st Rank A. Ivanov, "Great Britain's Naval Ports," *ZVO*, 3 (1978), p. 96.
72 S. Grishulin, "The Modernization of Great Britain's Air Defense Control System," *ZVO*, 5 (1985), p. 52.
73 Lt. Col.–Engineer V. Tamanskii, "Improving Great Britain's Ground Air Defense Weapons," *ZVO*, 11 (1980), p. 54.
74 Kolosov, *Voenno-politicheskii kurs*, p. 10.
75 Vorontsov, *SShA i Zapadnaya Evropa*, p. 227.
76 Stanislav Kondrashov, "Washington's Lap Dog or Its Own Position?" *Izvestiya*, April 24, 1986, p. 5.
77 V. F. Davydov, T. V. Oberemko, and A. I. Utkin, *SShA i zapadno-evropeiskie "tsentry sily" [U.S. and West European "Power Centers"]* (Moscow: Nauka, 1978), pp. 103–104.
78 See, for example, Vorontsov, *SShA i Zapadnaya Evropa*, p. 222.
79 Ibid., pp. 222–223.
80 Trukhanovskii, *Voprosy istorii*, 6 (1985), p. 35.
81 Kolosov, *Voenno-politicheskii kurs*, p. 60.
82 Ibid., p. 65.
83 Nikolaev, *ZVO*, 7 (1984), p. 10.
84 Kolosov, *Voenno-politicheskii kurs*, p. 65.
85 Golubev, *Voprosy istorii*, 7 (1984), p. 56.
86 Gennady Gerasimov, "London's Nucleophilia," *Sovetskaya kultura*, April 4, 1987, as translated in *SWB*, April 22, 1987, p. 9.
87 TASS, "The Conversation of M. S. Gorbachev and M. Thatcher," *Pravda*, December 8, 1987.
88 TASS, "M. S. Gorbachev Receives Geoffrey Howe," *Pravda*, February 17, 1988.
89 A. Lyutyi, "NATO Zigzags," *Pravda*, November 5, 1988.
90 Kolosov, *Voenno-politicheskii kurs*, p. 10.
91 Ibid., p. 104.
92 Ibid., p. 106.
93 Davydov et al., *SShA i zapadno-evropeiskie*, p. 105.
94 I. N. Shchelokova, *Problemy evropeiskoi bezopasnosti i politika Anglii [Problems of European Security and England's Policy]* (Moscow: Mezhdunarodnye otnosheniya, 1982), p. 31.
95 M. Belyaev, "Who Benefits from the Arms Race?" *KZ*, January 24, 1985, p. 3.
96 S. P. Madzoevskii and E. S. Khesin, "Great Britain in Today's World," *MEMO*, 8 (1980), p. 55.
97 Ibid., p. 56.
98 Kolosov, *Voenno-politicheskii kurs*, p. 178.
99 Ibid., p. 184.
100 Madzoevskii in *MEMO*, 4 (1975), p. 125.
101 Kolosov, in Kishilov, ed., *Zapadno-evropeiskaya integratsiya*, p. 246.
102 Ibid., p. 245.
103 Kolosov, *Voenno-politicheskii kurs*, chap. 1, especially p. 10.
104 G. V. Kolosov, "Military-Political Aspects of the West European Integration Process," *MEMO*, 4 (1987), pp. 34–42.
105 Kolosov, *Voenno-politicheskii kurs*, p. 140.

106 See, for example, Lt. Col. I. Gavrilov and Sr. Lt. S. Tomin, "Great Britain's Aeromissile Industry," *ZVO*, 7 (1985), pp. 51–57.
107 V. Babushkin, "A Fighter for Western Europe," *KZ*, July 27, 1988.
108 Kolosov, *Voenno-politicheskii kurs*, pp. 143–144.
109 A. Vladimirov, "The Export of British Arms," *ZVO*, 1 (1982), pp. 19–23.
110 Kolosov, *Voenno-politicheskii kurs*, pp. 190–191.
111 Yu. Kovalenko, "M. Thatcher's Two Trips," *Izvestiya*, November 23, 1986, p. 4.
112 "A Conversation of Examples," *Izvestiya*, April 28, 1987.
113 A. Lyutyi, "Preparing Compensation," *Pravda*, December 19, 1987.
114 See, for example, S. P. Peregudov, "Great Britain: The Elections Are Over, the Rivalry Continues," *MEMO*, 11 (1983), pp. 111–114.
115 Gorshkov, *SWB*, May 13, 1987, p. 7.
116 Arkadii Maslennikov, "Why the Tories Won," *Pravda*, June 13, 1987.
117 V. Kelin, "USSR-England: Experience in Cooperation in Combatting a Common Danger," *MEMO*, 8 (1984), p. 112.
118 Golubev, *Voprosy istorii*, 7 (1984), p. 57.
119 Ryzhikov, *Britanskii leiborizm*, p. 262.
120 Vorontsov, *SShA i Zapadnaya Evropa*, p. 228.
121 Maslennikov, *Pravda*, January 13, 1986, p. 5; V. S. Mikheev, "The Westland Affair in Various Dimensions," *Izvestiya*, February 18, 1986, p. 5.
122 Kolosov, *Voenno-politicheskii kurs*, pp. 238–239.
123 Roshchupkin, *KVS*, 5 (1985), p. 85.
124 TASS, "M. S. Gorbachev's Meeting with British Parliamentarians," *Pravda*, May 27, 1986, p. 1.
125 Trukhanovskii, *Voprosy istorii*, 6 (1985), pp. 36–37.
126 Golubev, *Voprosy istorii*, 7 (1984), p. 52.
127 Ibid., p. 55.
128 Ibid., p. 49.
129 Vladimir Bogachev, "Truly Balanced Approach Is Needed," *SWB*, March 6, 1987, p. 2.
130 Viktor Levin, "International Diary," as translated in *SWB*, March 23, 1987, p. 4.
131 Vladimir Chernyshev, "Prime Minister's Visit to Moscow," *SWB*, March 16, 1987, p. 1.
132 Neil Kinnock, as quoted by Vsevolod Shishkoviskii, "Mrs. Thatcher's Visit," *SWB*, April 3, 1987, p. 14.

7
A Nuclear-Free Zone in Northern Europe[1]

Susan L. Clark

The Soviet public diplomacy effort has been significantly revitalized under General Secretary Mikhail Gorbachev. Within this effort, the role of nuclear weapons has been of particular importance, as seen in the emphasis the Soviets have placed on developing a "third-zero" option in nuclear weapons and on the ultimate denuclearization of Europe. Such policies, and especially attempts to split the North Atlantic Treaty Organization's (NATO) nuclear powers from its nonnuclear countries, are clearly reflected in the attempts to establish a nuclear-free zone in Northern Europe. Soviet behavior with respect to this idea also illustrates the differences in the Soviet approach toward the large and small countries in Western Europe. The Soviets perhaps perceive greater opportunities for the success of such a zone proposal in the northern region than in other areas.

According to one statement in the authoritative military journal *Zarubezhnoe voennoe obozrenie*, the West plans

> to win a "decisive victory" in antisubmarine warfare and "shut up" the Soviet Navy in seas washing Scandinavia.... Some bloc leaders believe that if a war in Europe is not won on the northern flank it will be lost entirely. Such lines are made the basis of further integration of countries in NATO's nuclear infrastructure.... One cannot help but be alarmed

> by reports that in attempting to get around the INF Treaty, the North Atlantic Alliance is seeking methods of "compensating" for the loss of Pershings and ground-launched cruise missiles specifically on the northern axis by deploying sea-launched and air-launched cruise missiles in the North Atlantic.... In this situation the nonnuclear status of this region's countries can be lost even in peacetime.[2]

This, then, represents the current Soviet argument for the need to establish a nuclear-free zone in this region. Nevertheless, this concept has hitherto been undermined by the traditionally heavy-handed Soviet approach, as epitomized by Leonid Brezhnev's style of leadership. The new diplomatic style of Gorbachev obviates at least this problem, although other obstacles remain.

Formation of the Idea

The concept of establishing a nuclear-free zone to encompass the countries of Northern Europe has existed in the form of various initiatives for more than two decades but has yet to yield any substantive results. This is not to say, however, that the prospects for creating such a zone in the future are nil. Interest in this idea has revived during the past several years, sparked in part by NATO's deployment of intermediate-range nuclear forces (INF), revitalized peace movements in many countries, and some changes in the Soviet Union's position on this issue. It is the latter factor that is of the greatest interest to this analysis, serving to illustrate changing Soviet perceptions of the Nordic balance.

The Soviet Union has consistently supported the notion of a nuclear-free zone in Northern Europe and did, in fact, informally suggest such an idea back in the late 1950s. In this effort, the Soviet Union has found the Finnish government to be a willing collaborator. The earliest substantive proposal for a nuclear-free zone in Northern Europe was issued by Finland's President Urho Kaleva Kekkonen in May 1963. The key points contained in this proposal boil down to the following: that the states in this region would not possess or procure nuclear weapons and that they would not allow foreign nuclear weapons to be stationed on their soil (although it is not certain whether this latter stipulation was an absolute ban or merely a conditional one that would not apply under certain circumstances).[3] The main aim of the proposal was to stabilize the situation in this region by formalizing its extant nuclear-free status.

Reaction to this initiative among the other North European countries was mixed at best. Not surprisingly, NATO opposed the idea from the start, based on the argument that because the Nordic countries had no nuclear weapons and were opposed to acquiring them, a nuclear-free zone already existed in principle; there was no reason to formalize what already existed de facto. Specifically, both the Danish and Norwegian governments rejected the proposal since they had already vowed not to base nuclear weapons on their soil in peacetime. As Soviet analysts argued, however, the two countries' rejection of the idea was designed to satisfy their NATO allies, who wanted to leave themselves a "nuclear loophole" in this region. That is, NATO wanted to be able to station nuclear weapons there in a crisis situation.[4]

The Soviet Union, on the other hand, has consistently supported nuclear-free-zone initiatives, occasionally even modifying its own position to meet some of the concerns of the North European countries. For instance, whereas in 1959 the Soviets encouraged Norway and Denmark to accept the idea of a nuclear-free zone as the first step toward their neutrality, they have since dropped this concept in favor of the argument that such a zone would not entail any changes in the security policies of these countries.[5] But while certain aspects of Soviet support for the idea may have changed over time, the reasons underlying this support have not. First and foremost, the Soviets seek to use this idea as a means of weakening the defense capabilities of NATO's northern flank in general and as an instrument to weaken the ties between these North European members of NATO and their other NATO allies. In addition, beginning in the late 1970s the concept was also used to encourage North European opposition to U.S. INF deployments in Europe. As Orlan Berner explains, Soviet interest in inducing Norway and Denmark to sign a binding commitment "can now be assumed to be more political than strictly military." If these two countries were to become part of a formalized nuclear-free zone, it would rule out the nuclear option for them in wartime, would weaken the ties between them and the rest of NATO, and could provide impetus to other antinuclear forces throughout Western Europe.[6]

The 1970s: More Proposals but Little Progress

Throughout the 1970s Finland—with constant support from the USSR—advocated adoption of the nuclear-free zone concept, frequently using speeches in the United Nations as the forum for advancing this idea. There are, however, differences in emphasis between the proposals from the early 1970s and those advanced later in the decade. During the first half of the 1970s, the Finnish initiatives were colored by the prevailing detente atmosphere and, in this connection, emphasis was placed on international linkages whereby the Nordic countries would not be separated from the rest of Europe, but rather a North European nuclear-free zone would become "part of broader European arrangements." This was to be accomplished by linking nuclear-free-zone talks with Europe-wide arms control negotiations.[7] In contrast, President Kekkonen's May 1978 proposal illustrated the change in emphasis that emerged in the late 1970s owing to the cruise missile debate.

In 1974 the United Nations General Assembly approved Finland's initiative "to undertake comprehensive research into the problems of a nuclear-free zone, the creation of which would be a serious obstacle to proliferating nuclear weapons."[8] In October 1974, Chairman of the Presidium of the USSR Supreme Soviet N. V. Podgornii traveled to Finland to express the USSR's support for Kekkonen's proposal and stated that the Soviet Union was prepared to act as a guarantor, along with the other nuclear powers, of the region's nuclear-free status.

The detente atmosphere contributed to optimism about the possible success of such an initiative, as illustrated in Kekkonen's March 1975 declaration that detente had created the preconditions for fulfilling the idea of a nuclear-free zone. Two leading Soviet specialists on Northern Europe echoed these sentiments in a 1976 book, asserting that "although there are still influential circles in the Scandinavian countries who declare Finland's proposal to be 'unrealistic' and 'unnecessary,' today this discussion—in contrast to the 1960s—has turned into a concrete and realistic discussion of the problems connected with this proposal."[9] In addition, Soviet analysts believed that detente could be further developed through a nuclear-free-zone initiative; according to Yurii Denisov, the nuclear-free-zone idea "would surely meet the interests of the countries in that part of Europe [and] . . . would also add to the efforts to supplement political detente in Europe with military detente."[10]

But the emergence of new technologies raised dilemmas for the future of a nuclear-free zone in Northern Europe, a problem that President Kekkonen alluded to in his address to the Swedish Institute of International Affairs in Stockholm in May 1978. Singling out cruise missiles and neutron weapons in particular, Kekkonen argued that "the new military hardware facilitates the introduction of new elements of uncertainty into the policy of security in the North," mainly because these weapons could easily involve in a conflict states that had not previously been involved.[11] The primary concern was that American cruise missiles would likely have to fly through the air space of neutral countries in order to reach Soviet targets. Suggesting the use of his 1963 proposal as a starting point, Kekkonen unveiled a three-part plan: to develop a treaty on arms control in the Nordic region, to develop a treaty strengthening the nuclear-free status of Northern Europe, and to establish a guarantee by the nuclear powers that nuclear weapons would not be used against the states belonging to this zone. The aim of Nordic negotiations should be, he continued, "the formation of a separate treaty system that would most completely isolate the Nordic Countries as a whole from the influence of nuclear strategy and, especially, from the latest nuclear arms."[12]

Reaction among the North European countries to this proposal was again mixed, with the greatest opposition coming from NATO countries that still wanted to protect their right to use nuclear weapons in a crisis situation. As before, the arguments against such a plan focused on the facts that there are no nuclear weapons in the region, all the North European countries have already signed the nuclear nonproliferation treaty, such an agreement could change the balance of power in the region in the Soviet Union's favor, and such an agreement would, in any case, hold only in peacetime.[13]

As was to be expected, the Soviet Union supported this round of proposals as well, although it was quick to point out the impossibility of including its own territory in such an arrangement. In September 1978, Alexei Kosygin provided the first authoritative Soviet statement on the new initiative: "The establishment of a nuclear-free zone in Northern Europe would promote stability in that part of the world and help to strengthen the security of all the European countries. This measure could also do a great deal to solve the pressing problem of ensuring non-proliferation of nuclear weapons in Europe and the world."[14] However, this nuclear-free-zone initiative contained two components to which the Soviets objected: the inclusion of Soviet territory and of seas, specifically the Baltic, in such a zone as well. In an important article in the authoritative journal *Mirovaya ekonomika i mezhdunarodnye otnosheniya* in March 1979, Yurii Komissarov argued that "the USSR is a nuclear power and, consequently, neither its territory nor any portion of it can be included in a nuclear-free zone created by nonnuclear states, or in a so-called 'security belt' adjacent to a nuclear-free zone; nor can the stipulations of a nuclear-free zone act as an obstacle for the navigation of Soviet naval ships through the Baltic Straits regardless of the types of weapons on board."[15] In addition, his argument continues, "It must also be recognized that proposals for the inclusion of a part of Soviet territory in the nuclear-weapon-free zone in Northern Europe and the removal of certain nuclear weapons and missiles from that part cannot be of practical significance under present circumstances, taking the development of nuclear technology into consideration. It would not constitute any real guarantee for the security of the Nordic countries which declare their territory a nuclear-weapon-free zone."[16]

Thus, while the Soviet Union was willing to support such initiatives and even to act as a guarantor of a nuclear-free zone, the inclusion of its own territory was out of the question; its status as a world power meant that any such unilateral steps would affect the entire international balance, not just the regional balance. This would remain the oft-quoted Soviet position until the beginning of the 1980s. While this stance did not make the job of selling the nuclear-free-zone idea any easier, in the 1980s the nuclear-free-zone movement was aided by the increased popularity of peace movements throughout Europe, and particularly in the Nordic region. Moreover, the Soviet Union also changed some of its own positions on the issue, which may have helped the cause even more.

The 1980s and Soviet Initiatives

The 1980s witnessed a significant revival of interest in the idea of a nuclear-free zone at both the governmental and public levels. The annual meeting of North European foreign affairs ministers in 1981 put this topic on their agenda for the first time, and it has remained on the agenda ever since. As Komissarov states, unlike in the 1960s, during this decade "not a single government of the Northern European countries is opposed to discussing [the idea of a nuclear-free zone]. Moreover, not only neutral Sweden, but also Norway, Denmark and Iceland have officially declared their positive attitude, in principle, to Finland's proposal."[17] Understandably, it is the socialist and other left-wing parties that support the concept more than do the conservative parties in these countries. But, as John Ausland indicates, the nuclear-free-zone proposal "has... become such an important outlet for public anxiety that even nonsocialist governments cannot oppose it directly."[18]

According to Soviet analysts, a denuclearized zone would ensure the security of the countries in that zone, would not alter the balance of power, and would encourage the nonproliferation of nuclear weapons.[19] In addition, they are quick to point out the high degree of public support for this initiative among the Nordic countries' populace. In fact, they argue, the U.S. and NATO leaderships are primarily responsible for this initiative's lack of success in Northern Europe. According to Western reasoning, if such an idea were enshrined in treaty form, it would undermine NATO's flexible response strategy and the concept of nuclear deterrence. In addition, the main argument used against the nuclear-free-zone concept is the fact that the area is already, in essence, a nuclear-free zone. There is no evidence that this status would change (at least in peacetime), and there is therefore no need to formalize the arrangement.

The reasons for the increased interest in this idea revolve around the heightened nuclear debate within Europe (particularly over the cruise missile), the increased emphasis placed on the strategic importance of the northern flank by U.S. and NATO plans, changes in the Soviet positions vis-à-vis the inclusion of Soviet territory (1981 and 1986) and the Baltic Sea (1983) in such arrangements, and changes in the attitudes of the Norwegian and Danish Labor parties toward this concept.

Soviet analysts have concentrated considerable attention on the impact that cruise missile technology has had on the security of this region; namely, this technology has "radically changed the situation."[20] Representative of this analysis is a 1985 article by M. Kostikov appearing in *Pravda*: "The appearance of new American nuclear missiles in direct proximity to the Northern European countries gives real, evil characteristics to the threat of using their air space when launching these weapons, aimed against the USSR and its allies.... The flight trajectory of these missiles, whether land-based, for example in England, or sea-based in the Norwegian Sea, will pass over the territories of not only the Scandinavian members of NATO, but also over Sweden and Finland."[21] It is, the Soviet argument continues, partly because of the cruise missile deployments that the nuclear-free-zone concept "is winning the support of those political forces which only a few years ago brushed it aside."[22] And even following the signing of the INF agreement, interest in the idea has not completely faded.

In general, the current Soviet position on a North European nuclear-free zone is that it will act as a guarantor of the zone's nuclear-free status (even if other nuclear powers refuse to do so); it will sign multilateral or bilateral treaties with the participating countries, pledging not to use nuclear weapons against these countries; it will discuss extending the nuclear-free zone to the Baltic Sea; and it will consider certain "substantial" measures pertaining to its own territory (see below). The Soviets argue that working to create such a zone increases stability in the region, and they frequently link this idea with enhancing the security of Europe as a whole. For example, as Savanin stated in *Krasnaya zvezda*, "our country has repeatedly affirmed its adherence to the idea of creating a nuclear-free zone in this region, believing that such zones are a realistic way to a nuclear-free Europe."[23]

According to Soviet analysts, the main obstacle to the realization of this idea has been opposition to it by the United States and NATO. The United States has argued that such an agreement would contradict its nuclear commitments to its European allies. In this connection, one Soviet analyst argued that the Danish and Norwegian policies against allowing nuclear weapons on their soil in peacetime "is being increasingly emasculated as the U.S. and NATO militaristic policies are intensified.... The Nordic NATO countries are increasingly involved in the U.S. nuclear strategy."[24]

Inclusion of Soviet Territory

A main point of contention in discussions about the desirability and feasibility of establishing a nuclear-free zone has long been the question of including some part of Soviet territory in such an agreement. In the 1960s, Soviet analysts stated that this notion was "senseless," particularly because Soviet weapons—especially its intercontinental ballistic missiles (ICBMs)—could reach any point on earth.[25] As indicated earlier, another argument against inclusion of Soviet territory was the fact that the Soviet Union is a world nuclear power and any concessions it might make on this score would affect the international balance of power, not just the regional balance. Such reasoning persisted until June 1981, when Leonid Brezhnev indicated for the first time that a shift in position was in the making. In addition to stating that the Soviet Union would not use nuclear weapons against any North European countries belonging to a nuclear-free zone, the general secretary declared that "the Soviet Union . . . did not rule out the possibility of considering some other measures pertaining to its own territory in the area bordering on a nuclear-free zone in the North of Europe."[26] Many people in the West speculated about what these measures might be. In November 1981, Brezhnev further stated that these measures could be "substantial."

For several years the Soviets refused to specify the substance of these measures. A 1984 book by L. S. Voronkov explains this Soviet reluctance:

> The measures the Soviet Union is willing to consider in those portions of the country bordering on the Nordic countries will depend on a number of questions, such as which countries are included in the zone and the conditions under which the zone is formed. In addition, they would also depend on whether or not the West would provide guarantees for the zone or if the Soviet Union would be the sole guarantor. For these reasons, the Soviet Union is unwilling to state the specific concession it would make at this stage.[27]

It would not, in fact, be until November 1986—a full five years later—that the Soviet Union would actually detail the substance of this pledge.

It is striking that the best-known and most authoritative Soviet source on the northern flank, Yurii Komissarov, stated only four months before the major policy change that Soviet forces, namely those on the Kola Peninsula, "have never been of a regional focus and have not nor do not threaten the security of the Northern European countries, but rather are called upon by the need to reestablish military-strategic parity on a global scale that has been disrupted by the U.S. and NATO."[28] He also previously asserted that "impracticable demands are being made with regard to including the north-western regions of the USSR in the zone."[29] Thus, even as well-informed a source as Komissarov appeared to be unsuspecting of the impending November 1986 initiatives. The timing of the initiatives may have been determined primarily by the Reykjavik summit.

On November 13, 1986, E. K. Ligachev, Secretary of the Communist Party of the Soviet Union (CPSU) Central Committee, delivered a speech in Helsinki outlining four basic points of the Soviet position vis-à-vis a North European nuclear-free zone. In it he declared:

> 1. We have already dismantled medium-range missile launchers on the Kola Peninsula and a large portion of the launchers for such missiles in the remaining area of the Leningrad and Baltic military districts and have moved

several divisions of operational-tactical missiles from these districts. This is a *concrete confirmation* of the USSR's earlier *declared readiness to consider certain, and substantial measures with respect to its own territory adjacent to a future nuclear-free zone.* [Emphasis added.]

2. In confirming our support for the idea of according the Baltic Sea nuclear-free status within the framework of realizing the proposal for a nuclear-free North, we could, in the event an agreement were reached on this issue among the appropriate states, withdraw ballistic-missile submarines from the Soviet Baltic fleet.

3. In supporting the idea of possible confidence-building measures in Northern Europe and the adjoining North, Norwegian, Barents, and Baltic Seas, the Soviet Union proposes starting to limit the intensity of major exercises in this region. Such exercises—of 25,000 men and more—should not be conducted more than once or twice every one or two years.

4. In order to strengthen naval confidence-building measures, we share the idea of using the positive experience from the 1972 Soviet-American agreement to prevent incidents in the open sea and in the air spaces over them. As is known, the Soviet Union signed a similar agreement with England in July 1986.[30]

According to one Soviet analyst, these initiatives set forth under the new Gorbachev leadership "demonstrate the Soviet Union's new thinking, aimed in deeds and not in words at changing the dangerous course of events in the world and in individual regions."[31]

Aside from the propaganda benefits to be derived from this move, particularly in the wake of Reykjavik, the Soviets were striving to breathe new life into their efforts to create a nuclear-free world, including in this region. In addition, the initiative appeared to be timed to a rise in interest about this issue among some of the Nordic countries, particularly Norway.[32] A joint Soviet-Finnish communiqué of January 1987 provided hints that there might be more to come, since the Soviets again indicated their willingness to act as a guarantor of a nuclear-free zone and "to examine the question of certain steps relative to its own territory" in addition to discussing the granting of nuclear-free status to the Baltic Sea.[33]

A Nuclear-Free Baltic Sea?

Another major stumbling block to any significant breakthrough in establishing a nuclear-free zone in Northern Europe has been the status of the Baltic Sea in these proposals. The following statements illustrate the change in Soviet policy on this issue.

First, in *Nuclear-Free Status for Northern Europe*, L. S. Voronkov reasons that the Baltic should not be included in a Northern European nuclear-free zone because the Baltic is part of the overall military balance in Europe, not just in the Nordic area. If the Baltic is to be included in these discussions, then other waters touching the Scandinavian countries should be as well. The Baltic's status should, in fact, be resolved within the context of a broader picture, as part of a zone extending throughout Europe to the Mediterranean.[34]

But in June 1983 General Secretary Yurii Andropov delivered a speech honoring Finnish President Mauno Koivisto, in which he affirmed that the proposal for a Nordic nuclear-free zone could be extended to include the Baltic Sea. As indicated earlier, in Ligachev's November 1986 speech the Soviets also indicated the possibility of withdrawing their ballistic-missile submarines from Baltic waters if an agreement were reached. In addition, Yurii Komissarov rejected another of Voronkov's tenets when he "dismisses assertions that the Soviet Union would only be prepared to discuss nuclear-free status for the Baltic in a broader European context. He says that a nuclear-free Baltic cannot be brought about simply by snapping one's fingers, but that a whole series of special questions exist. For example, in connection with the drawing of demarcation lines in the Baltic."[35] Komissarov reasons that these are issues that can be resolved at the negotiating table and should not impede participants from ever reaching the table.

Gorbachev's Murmansk Speech

On October 1, 1987, Gorbachev delivered a major speech in Murmansk in which he unveiled a six-point proposal for the northern region encompassing cooperation on environmental protection, scientific research, limiting military activities, and so on. Specifically, his speech did the following: (1) Called for the creation of a nuclear-free zone in Northern Europe. In this context, Gorbachev reiterated the Soviet Union's willingness to guarantee the zone and to implement measures with respect to its own territory, citing the dismantling of the missile launchers on the Kola Peninsula and nearby regions as an example of one step the USSR had already made unilaterally. (2) Praised Finnish President Koivisto's 1986 proposal to limit military operations at sea in the North European area. (3) Stated that the Soviet Union places great value on peaceful cooperation aimed at using the natural resources in the polar region. (4) Called for a conference to be held in 1988 that would involve all nations with authority over the polar regions in order to coordinate their scientific research. (5) Advocated cooperation in environmental protection in this region. (6) Suggested that Soviet icebreakers could keep open the seaways north of the Soviet Union between Europe and the Pacific for foreign commercial ships, provided the state of international relations so permitted.

The greatest amount of attention has been focused on those sections of the Murmansk speech that address security issues and are drawn from the Koivisto proposal. In addition to the familiar proposal for a nuclear-free zone, the idea of limiting naval activities encompassed restricting both naval and air operations in the Baltic, North, and Greenland seas. It also called for extending confidence-building measures to the region, providing notice of major naval and air force exercises, and banning naval activities in mutually agreed zones, such as international straits and busy shipping lanes. In addition, in January 1988 Ryzhkov again suggested that the six Golf-class nuclear-armed submarines might be withdrawn from the Baltic fleet as one possible measure the Soviet Union could adopt with respect to its own territory.[36]

To all accounts the main motivation behind Gorbachev's initiative was linked to the increasing prospects for a U.S.-Soviet agreement to eliminate their INF systems. According to Soviet analysts, even before the INF agreement was signed, the West was expressing concerns about whether such an agreement would negatively affect the northern region and exacerbate tensions there. Because of this, the Soviets allege, NATO began looking for ways to compensate for the elimination of these missiles, including a plan "to increase the arsenal of cruise missiles, to build up the combat squadron in the North Atlantic and the Norwegian Seas, and to create a shore infrastructure of the U.S. Navy on the northern flank."[37] The linkage between Gorbachev's Murmansk initiative and the INF treaty is made quite clear in the following TASS statement: "The Soviet Union supports Finland's [October 1986] proposals aimed at strengthening confidence and reducing military activities in the Northern region. These initiatives are especially important now, when the issue is one ... of an unhealthy interest in this region that is strategically beneficial for implementing the 'forward borders' concept and for 'compensating' for medium- and short-range missiles abolished under the Soviet-American treaty."[38] A frequently quoted statement by Colonel-General N. F. Chervov further cautions that "we have gotten used to thinking that the risk of a war can arise on land, but it is not ruled out that it will arise at sea," particularly given the strength of U.S. and NATO naval forces.[39]

U.S. criticism of the Murmansk proposal has, in fact, been focused on the notion that agreeing to such restrictions on naval and air forces in the northern region would impede the West's reinforcement capabilities and weaken deterrence.[40] Reaction among the Nordic countries has tended to follow the policy lines already delineated in response to previous proposals in this region. What is worth noting is the fact that, although the Norwegian and Danish governments have been careful to stress the need for caution in appraising Gorbachev's proposal, they have been hesitant about rejecting it out of hand. For instance, an editorial in *Berlingske Tidende* notes that the substance may not have been new, but the tone was: "It was marked by the spirit of reconciliation that at present marks the relations between the East and the West.[41] In the same vein, Norway's Prime Minister Gro Harlem Brundtland voiced the opinion that the Murmansk initiative was "a new sign of the Soviet desire to enter into committed international cooperation even in a region and in areas which affected sensitive national interests.... Something that is positive about Gorbachev's speech is that it is yet another example of a more open and active approach to weighing interests against each other in order to reach agreement."[42]

One way in which the Soviets can continue to undermine Western cohesion, especially on the northern flank, would be through further reductions of their naval operations in these seas. For example, naval exercises with major surface ships in the Norwegian Sea declined from a peak of 456 sea days in 1985 to 207 in 1986 and 114 in 1987. Similarly, in the Barents Sea it is estimated that there has been a 50 percent reduction in these exercises.[43] The reasons for this reduced presence are many and varied (economic, military, and diplomatic), but one of its results may be the emergence of the perception among broad segments of the Nordic populace that the threat from the Soviet Union has been curtailed significantly.

Scandinavian Initiatives

During the 1980s, some of the North European countries themselves took initiatives to promote the nuclear-free-zone concept. Within multinational forums there have been several high-level meetings as well as conferences on the subject, albeit without many substantive results. On the national level, during the 1980s the parliaments in the Netherlands, Sweden, and Denmark passed resolutions obligating their governments to facilitate plans for turning Northern Europe into a nuclear-free zone.[44] In 1980, spurred by a speech by a then-senior official in the Norwegian Foreign Office, Jens Evensen, supporting the creation of such a zone, the Norwegian Labor party adopted a resolution calling for a North European nuclear-free zone in the context of a wider European agreement. And when this party came to power in May 1986, it called for stepping up efforts to establish such a zone. Along with the cruise missile deployment decision, this speech instigated a new round of discussions about the desirability and feasibility of such a proposal. But the Norwegian, Danish, and Swedish governments continued to insist that such a zone should be created only in the context of a wider European agreement.[45]

In the early 1980s, forces in Norway were increasing their support for the nuclear-free-zone idea. The United States reacted by warning that the consequences of such a move could be a U.S. refusal to assist Norway in a crisis situation. Since then, Norway's policies have received generally negative reviews from the Soviet Union, primarily because it believes that "Norway is gradually giving in to NATO pressure as the Americans push a new offensive strategy on their Northern flank."[46] Norwegian officials certainly do not deny their closer ties with NATO, but they explain that this enhanced cooperation is due primarily to the Soviet buildup in the north—especially its missiles on the Kola Peninsula. There is a real dichotomy within Norway on defense issues; serious doubts about nuclear weapons are coupled with the desire to remain a member of NATO. Reflective of the general Norwegian stance on this issue is the Colding report, issued in November 1985, which asserted that a nuclear-free zone must be resolved only in a Europe-wide context and should be linked to the results of the disarmament talks in Europe. Additionally, the report suggested that the northwestern areas of the USSR, namely the Kola Peninsula and the Baltic Sea, should be included, and it made demands for conventional arms limitations in the region owing to Soviet conventional superiority.[47] In early 1986 there was no ambiguity in the Soviet reaction to this idea. According to Komissarov, "The Soviet Union will not accept Norwegian demands for limitations on the numbers of nuclear and conventional weapons on the Kola Peninsula, which are a precondition for a nuclear-free zone in the Nordic area."[48] Yet in November 1986 the Soviets did subsequently compromise on inclusion of this very territory.

Denmark, like Norway, generally adheres to the position that a nuclear-free-zone agreement could be achieved only as a part or a consequence of a broader East-West agreement. While neither country opposes discussing this issue, they do oppose granting these talks negotiating status. Not surprisingly, the socialist parties in both countries (as well as the Soviets) have criticized the governments on these positions. The bottom line for these two countries is simply that if they want to remain in NATO, they will have to receive U.S. and NATO approval for any such agreement.

Previous Danish unity on its security policy has disappeared, and a special *Folketing* committee was even established to try to find some common ground between the liberal government and the Social Democrats. The government remains skeptical of a denuclearized zone, whereas the Social Democrats strongly favor such an idea. The latter argue that Soviet statements about wanting to relax tension must be taken seriously and that there has been a substantive change in Soviet policies; the government remains unconvinced. The Soviets seem to hold out little hope of changing the Danish government's mind, as evidenced in their failure even to discuss the zone idea with the prime minister during an October 1986 visit.[49] Nonetheless, in April 1988 Denmark's Conservative-led government suffered a defeat in parliament when the social democratic opposition succeeded in passing a resolution that the Danish government must remind visiting warships about the prohibition of nuclear weapons on Danish territory in peacetime. The nuclear issue thus remains a controversial subject among the Danish populace, a fact the Soviets might yet be able to turn to their advantage.

During his tenure, Swedish Prime Minister Olaf Palme consistently supported the nuclear-free-zone idea, and his successor, Ingvar Carlsson, has underscored that "the government will continue to actively advocate creating a nuclear-free zone in Northern Europe."[50] Certainly the deployment of cruise missiles has served only to increase Swedish support for these proposals, viewing as it does the penetration of its air space by such missiles to be a violation of its neutrality.[51]

One Finnish initiative in the mid-1980s that received a cool reaction from the Soviet leadership (at least initially) was the idea of establishing confidence-building measures to limit naval activities. According to one Soviet analyst, the idea set forth in October 1986 received a "positive reaction" from Sweden, Norway, and other Scandinavian countries. For the Soviet Union's part, it "understood the constructiveness of this proposal" and "supports" it,[52] although this support did not appear to be as enthusiastic as had been the case for previous Finnish security concepts. According to one Finnish commentary examining other countries' reactions to this initiative, for the Soviet government "right now [October 1986], ... issues concerning outer space have clear priority in Soviet policymaking."[53] In fact, not until Gorbachev's Murmansk speech one year later, in October 1987, did the Soviets appear to support this Koivisto proposal fully.

Prospects for the Future

Calls for establishing a nuclear-free Northern Europe have not met with much success during the past 20-odd years, and it is unclear whether the future holds any brighter prospects. There is no doubt that the issue has received increased visibility in the past few years. It is now always on the agenda of meetings of the North European foreign affairs ministers. In November 1985 the first high-level conference of North European countries met in Copenhagen to address this issue. The prime ministers of the five countries met in August 1986 in Denmark to discuss the idea, and a parliamentary commission from these countries was established and released a draft treaty as well as a watered-down final report. Thus, even if there was no concrete result, the level and extent of talks certainly rose during the 1980s.

The Soviets argue that if Norway and Denmark adopt more realistic policies, the chances for a nuclear-free-zone agreement would be much greater.[54] Yet many in the West doubt the Soviet Union's genuine interest in such an agreement. It is quite possible that the Soviets see real benefits to be derived simply by keeping the idea alive, rather than by actually signing a treaty. As it stands, they have made a visible gesture by withdrawing forces from the Kola Peninsula, but they are not bound by any legal document to continue such a commitment in the future. Thus, it may be that "the discussion of and the activity for such a [nuclear-free-zone] plan in the Nordic countries, rather than its achievement, is the Soviet Union's primary objective."[55]

Fortunately for the sake of Western security interests, it seems unlikely that there would simultaneously be governments in both Norway and Denmark that would advocate membership in a nuclear-free zone. What cannot be foreseen, however, is the impact of possible additional Soviet initiatives in this area. It is clear that the USSR has become much more effective in its dealings with the West, mainly through the use of effective propaganda. Certainly given Gorbachev's diplomatic skills, the carrot of adopting substantial measures affecting Soviet territory itself will continue to be held out in the hope of swaying Western opinion. With the signing of an INF agreement between the Soviet Union and the United States, the likelihood of increased pressure for a North European nuclear-free zone is that much greater. Thus far the Soviets have dismantled missile launchers on the Kola Peninsula and have indicated their willingness to eliminate the Golf-class nuclear submarines in their Baltic fleet. The West must be prepared for similar initiatives in the future. Western reaction to his Murmansk speech may not have met Gorbachev's expectations, but this hardly means that such proposals will disappear from the Soviet agenda. On the contrary.

If implemented, a nuclear-free zone in Northern Europe would almost certainly damage NATO unity, mainly by creating a split between this region and the rest of NATO, and would reduce the effectiveness of the coalition's deterrent strategy. In addition, as Robert German has pointed out, if such an agreement were signed, with the Soviet Union acting as one of its guarantors, the Soviets could then "no doubt claim a legitimate right to 'oversee' Nordic defense activities in the name of verification."[56] While the West must remain unified in opposing such moves that would undermine security, it could further explore other Soviet proposals, such as cooperation on environmental protection in the northern region. Progress on these fronts would allow the West to project a more positive image rather than one that generally rejects Soviet proposals out of hand.

The West cannot afford to ignore any opportunities to apply pressure to the Soviets in order to obtain additional concessions. There have been changes in the Soviet position on the nuclear-free-zone issue; the fact that these changes have occurred should not be overlooked. It should not be ruled out that the West might be able to obtain additional benefits without having to make concessions of its own, provided it stands firm and unified.

Notes

1. Research for this chapter was completed in August 1988.
2. N. Alekseev, "Nonnuclear Zones: Important Factor of European Security," *ZVO*, 5 (1988); translated in JPRS-UFM-88-011, p. 6.
3. Osmo Apunen, "Nuclear-Weapon-Free Areas, Zones of Peace and Nordic Security," *The Yearbook of Finnish Foreign Policy, 1978* (Finnish Institute of International Affairs, 1979), pp. 2–4.
4. Yu. I. Goloshubov, *Skandinaviya i evropeiskaya bezopasnost' [Scandinavia and European Security]* (Moscow: Mezhdunarodnye otnosheniya, 1971), p. 64. See also N. Neiland, "For a Nuclear-Free Northern Europe," *IA*, 6 (1983), p. 99.
5. Robert German, "Nuclear-Free Zones: Norwegian Interest, Soviet Encouragement," *ORBIS* (Summer 1982), p. 452.
6. Orlan Berner, *Soviet Policies toward the Nordic Countries* (Lanham, NY: University Press of America, 1986), p. 121.
7. Apunen, *Finnish Foreign Policy*, p. 10.
8. L. S. Voronkov, "Soviet and Finnish Scholars: A Joint Study," *IA*, 7 (1978), p. 129.
9. T. Barten'ev and Yu. Komissarov, *Tridtsat' let dobrososedstva [Thirty Years of Good Neighborliness]* (Moscow: Mezhdunarodnye otnosheniya, 1976), p. 213.
10. Yurii Denisov, "USSR-Finland: New Prospects for Cooperation," *IA*, 8 (1977), p. 87.
11. Neiland, *IA*, 6 (1983), p. 99.
12. As quoted in Yurii Denisov, "USSR-Finland: In the Spirit of the Helsinki Accord," *IA*, 3 (1979), pp. 34–35.
13. Yu. Komissarov, "Nuclear-Free Status for Northern Europe," *MEMO*, 3 (1979), pp. 110–111.
14. As quoted in Denisov, *IA*, 3 (1979), p. 35.
15. Komissarov, *MEMO*, 3 (1979), p. 114.
16. Yu. Komissarov, "The Future of a Nuclear-Weapon-Free Zone in Northern Europe," in *The Yearbook of Finnish Foreign Policy, 1978* (Finnish Institute of International Affairs, 1979), p. 30.
17. Yu. Komissarov, "Two Approaches to Security Problems in Northern Europe," *MEMO*, 7 (1986), p. 17.
18. John C. Ausland, *Nordic Security and the Great Powers* (Boulder, CO, and London: Westview Press, 1986), p. 56.
19. See, for example, V. Molchanov, "One Way towards a Nuclear-Free Europe," *New Times*, 10 (1984), p. 24.
20. Neiland, *IA*, 6 (1983), p. 100.
21. M. Kostikov, "Nuclear-Free Status for Northern Europe," *Pravda*, February 22, 1985. See also D. Pogorzhel'skii, "Nuclear-Free Status for Northern Europe," *KZ*, March 20, 1985.
22. L. Vidyasova, "Nordic Countries Fighting for Security," *IA*, 1 (1985), p. 83. See also Neiland, *IA*, 6 (1983), p. 99.
23. L. Savanin, "In a Distorting Mirror," *KZ*, November 7, 1986.
24. Vidyasova, *IA*, 1 (1985), p. 83.
25. V. Prokofev, *Severnaya Evropa i mir [Northern Europe and Peace]* (Moscow: Mezhdunarodnye otnosheniya, 1966), p. 136.
26. Neiland, *IA*, 6 (1983), p. 102.
27. As quoted in Jan-Magnus Jansson, "Editorial," *Hufvudstadsbladet*, October 10, 1984; translated in JPRS-WER-84-139, p. 3.
28. Komissarov, *MEMO*, 7 (1986), p. 19.
29. Yu. Komissarov, "Problems of Peace and Security in Northern Europe," *IA*, 7 (1985), p. 67.

30 E. K. Ligachev in TASS, "Press Conference in Helsinki," *Pravda*, November 14, 1986.
31 A. Sychev, "A Nuclear-Free Zone Is Possible," *Izvestiya*, November 16, 1986.
32 "Soviets Push Nuclear-Free Northern Europe," *Washington Post*, November 15, 1986, p. A29.
33 TASS, "Soviet-Finnish Communiqué," *Pravda*, January 10, 1987.
34 L. S. Voronkov, *Severnoi Evropa—bez'yadernyi status [Nuclear-Free Status for Northern Europe]* (Moscow: Nauka, 1984), pp. 100–101. It should be noted that this book went to the printer in February 1983, even though it was not published until 1984. This could explain the differences between this text and the speech Andropov made in June 1983.
35 Stefan Lundberg, "The Soviet Union Rejects Norwegian Conditions for a Nuclear-Free Zone," *Dagens Nyheter*, February 20, 1986; translated in *FBIS*, February 28, 1986, p. G3.
36 TASS, "Northern Europe: The Path to Security," *Pravda*, January 23, 1988.
37 N. Ivanov, "Atlanticists' Intrigues," *Izvestiya*, December 19, 1987. See also Col. Gen. V. N. Lobov, "Northern Europe: To Strengthen Stability and Security," *KZ*, January 29, 1988.
38 TASS, "USSR-Finland: A Formula for Cooperation," *Pravda*, April 8, 1988.
39 TASS, "Lasting Peace for Northern Europe," *Izvestiya*, January 24, 1988.
40 See, for example, Admiral C. Trost, "Northern Seas Are Vital for NATO's Defense," *Long Island Newsday*, March 28, 1988, p. 50.
41 "The Variant of Murmansk," *Berlingske Tidende*, October 3, 1987; translated in JPRS-WER-87-085, p. 4.
42 Gro Harlem Brundtland, "A Desire for Cooperation," *Arbeiderbladet*; translated in JPRS-WER-87-200, p. 17.
43 Tonne Huitfeldt, "Major Drop in Soviet Norwegian Sea Exercises," *Jane's Defence Weekly*, March 5, 1988, p. 382.
44 "A Nuclear-Free North: Completely Realistic," *Izvestiya*, August 16, 1986.
45 Lindberg, in *Finnish Foreign Policy*, 1980, p. 29.
46 Berner, *Soviet Policies*, p. 161.
47 Komissarov, *MEMO*, 7 (1986), p. 18.
48 Lundberg, *FBIS*, February 28, 1986, pp. G2–3.
49 Michael Kuttner, "Nordic Nuclear-Free Zone of Little Interest," *Berlingske Tidende*, October 23, 1986; translated in *FBIS-WE*, October 28, 1986, p. P1.
50 "New Government," *Pravda*, March 14, 1986, p. 5.
51 S. Morgachev, "Security Questions in Northern Europe and Swedish Policy," *MEMO*, 8 (1985), p. 74.
52 L. Savanin, "For a Nuclear-Free North," *KZ*, December 9, 1986. Paavo Keisalo, "Koivisto's Sea Initiative: Will U.S. Attitude Defeat Entire Plan?" *Suomen Kuvalehti*, February 13, 1987; translated in JPRS-WER-87-085, p. 59.
53 Keisalo, JPRS-WER-87-085, p. 59.
54 Komissarov, *MEMO*, 7 (1986), p. 20.
55 Kuttner, *FBIS-WE*, October 28, 1986, p. P2.
56 German, *ORBIS*, p. 453.

8
Soviet Security Perspectives on Germany[1]

Hannes Adomeit
*Fletcher School of Law and Diplomacy,
Tufts University*

Problems of Analysis

The difficulty of this topic hardly needs to be emphasized, for how is one to arrive at an accurate view of Soviet security perspectives on Germany? First, it is evident that a broad spectrum of opinion has ranged from abstract and differentiated analysis to vicious anti-German propaganda. The spectrum has stretched from abstract, theoretical treatises on the role of West Germany in the international system to specific discussions of the weapons systems as well as of the tactical and operational doctrines of the *Bundeswehr* (the Federal Armed Forces). Some published opinion can clearly be recognized as part of the official line, as propagandistic and designed to influence West German security policies. Other opinion may be classified as semi-official, deviating from the established government and party line but doing so with official authorization in an effort to test the reactions abroad to possible changes in Soviet policy. Other published analysis may differ from standard interpretation without government backing and thus perhaps express the specific views of the author or a particular institution in the Soviet hierarchy.

There has, of course, always been some of this diversity in Soviet sources. But the trend from the end of World II to the present has been that published opinion has become richer and more attuned to reality. Today, *novoe myshlenie* (new thinking) and *glasnost'* are deliberately seen as contributing to a new political culture of conflict in the Soviet Union.

Consequently, more than before genuine discussion rather than centrally controlled propaganda can be found in print.

Second, Soviet security perspectives on Germany contain both constant and changing characteristics. There have been periods in which perceptions of heightened danger were clearly evident, as, for instance, during the Berlin crisis of 1948, West Germany's rearmament and its inclusion in NATO, the Berlin crisis of 1961, and to some extent, perhaps also at the time of the stationing of U.S. intermediate-range nuclear force (INF) systems on West German and other West European territory. But heightened awareness of danger has altered with perceptions of lower risk as, for example, after the building of the Berlin wall in 1961 and the subsequent consolidation of the German Democratic Republic (GDR), the changes in the military correlation of forces in favor of the Soviet Union in the 1960s, the evolution of detente and German Chancellor Willy Brandt's *neue Ostpolitik* (new policy toward the East). Despite the fluctuations, a trend has emerged: over time, Soviet threat perceptions have tended to become less acute and the Soviet leaders' confidence in their ability to manage security relations successfully in Europe has increased.

Third, there has been a mixture of defensive and offensive components in Soviet policies vis-à-vis West Germany. These have existed quite irrespective of Soviet military strategy toward Western Europe, which has been offensive throughout the postwar period. More specifically, successive Soviet leaderships have had different views concerning the feasibility of using military power to influence the domestic and foreign policies of West Germany. In the 1960s and 1970s, Soviet leaders perceived more opportunities for transforming military asymmetries into political leverage. The apex of such assumptions came in the late 1970s and early 1980s, in conjunction with the controversy over the stationing of INF missiles and the rise of the peace movement in West Germany and Western Europe. Since General Secretary Mikhail Gorbachev's ascendancy to power, however, there seems to be less confidence among Soviet policymakers in their ability to use nuclear and conventional preponderance in Europe for political purposes.

Fourth, in the Soviet view since 1917 security has not only been determined by political and military factors, it has also always had a significant ideological dimension. The legitimacy of the Soviet state and the Soviet Communist party rests on it, as does the legitimacy of the *pax Sovietica* in Eastern Europe. There, Marxist-Leninist ideology has had to contend—for the most part unsuccessfully—with the forces of nationalism and demands for autonomy, liberalization, decentralization, and democratization. More often than not, unrest and revolt in Eastern Europe have thus been both anti-Soviet (directed against the ruling party and the Marxist-Leninist system imposed by the Soviet Union) and anti-Russian.

This is especially important to bear in mind when considering the German problem, for because of the division of Germany, any manifestation of nationalism in East or West Germany is, in the Soviet view, potentially disruptive to the postwar order. If such a manifestation were to be coupled with a broad current of ideological revisionism in East Germany, the mixture would be even more explosive. Certainly some revisionist tendencies are always present in East Germany and it would be surprising if this were different. Given a common language and the common traditions and history of the two parts of Germany prior to World War II and an economically successful, socially dynamic, and politically pluralist West Germany, stability in East Germany often seemed to be tenuous and to some extent must still be regarded as such in Moscow. For these reasons it is quite appropriate to focus on Soviet security perspectives on Germany rather than on West Germany alone.

Soviet Security and the German Problem: The Postwar Era

Traditionally, security has always had, apart from such subjective factors as threat perceptions, fundamental objective territorial and military dimensions. But while it is true that in an era of nuclear weapons, intercontinental delivery systems, and global communications it is questionable whether division of the world into vital and peripheral geographic areas and spheres of influence makes sense, the transition from traditional to new forms of security requirements had barely begun in 1945. The full implications of the change became visible only at the time of the Cuban missile crisis. It is therefore quite understandable that in the postwar era, Joseph Stalin approached Soviet security from traditional perspectives. Furthermore, because Soviet leaders historically have defined security in ideological terms as well as politically and militarily, the Soviet Union has never really acted as a traditional nation state.

How, then, could the Soviet Union have achieved security in its various dimensions—ideological, political, socioeconomic, and military—vis-à-vis Germany in the postwar era? Probably by pursuing one of four broad courses of action: (1) a revolutionary transformation of the social and economic system of the whole of Germany under the leadership of a German Communist party controlled by Moscow, (2) a substantial weakening of the economic and military potential of Germany in conjunction with territorial reductions, (3) division or dismemberment and the continued long-term enforcement of this status by four-power control, and (4) the establishment of a unified, neutral Germany. Although these approaches were not mutually exclusive, each affected Soviet interests in a different way. As it turned out, each either failed—or seemed to fail—to provide satisfactory solutions to Soviet dilemmas.

"Revolutionary Transformation" of Germany

According to classic Marxist theory, prospects for a revolutionary transformation of Germany in the postwar period seemed bright. Capitalism in Germany had reached its "highest state"; it had become "overripe." Germany had, as Stalin commented, "an extremely qualified and numerous working class and technical intelligentsia."[2] Fascism and militarism, phenomena inherent in imperialism as Leninist doctrine held, had been defeated and discredited. The collapse of Nazi Germany had set in motion far-reaching processes of socio-economic change. The influx of several million immigrants and refugees from German regions formerly under Polish and Soviet control—Czechoslovakia and Hungary—as well as the large-scale destruction of housing, had created a fertile ground for social unrest. Finally, there was widespread feeling among the members of the two parties on the left, the Social Democratic Party (SPD) and the Communist Party of Germany (KPD), that its disunity had largely contributed to the rise of German fascism. Now broad alliance strategies appeared to be necessary to rebuild the German body politic.

All of these factors did not, however, suffice to turn the tide in the Soviet favor. The rape and plundering of the German civilian population, a punitive occupation policy, the large-scale dismantling of plants and requisitioning of products as reparations and, after 1946, the stifling of autonomous political, social, and economic activity in its zone of occupation combined to destroy any chances the Soviet Union may have had for spreading its system to all of Germany.

But from the beginning Stalin did not seem to place much faith in the realization of such a development. At various times he had revealed skepticism about achieving communist revolutions abroad without the direct support of Soviet power. As far as Germany was concerned, this was shown by his comments to Milovan Djilas, Yugoslav Communist party official and Cabinet minister, that "you cannot have a revolution [there] because you have to step on the lawn";[3] or when he is reported to have said to the Polish leader Stanislaw Mikolajczyk that communism fitted Germany "as the saddle a cow."[4] Thus the first option for achieving security did not seem realistic.

Emasculation of Germany

Concerning the second possibility—weakening Germany's economic and military potential and reducing its territorial base—several facets must be distinguished. As regards territorial questions, it was agreed at the Tehran conference that the northern part of East Prussia would be transferred to the Soviet Union and a strong Poland would be created with substantial territorial compensations in the north and west. Polish sovereignty was extended de facto to the River Neisse, Stalin thereby laying the basis for the most probable development: Polish-German hostility and Polish dependency on the Soviet Union. By 1948 events seemed to have drifted precisely in that direction. In addition to the probable political consequences, the territorial changes in east central Europe considerably weakened the power base of Germany and strengthened that of the Soviet Union (and Poland) at Germany's expense.

Soviet reparations policy was meant to serve similar purposes. Of course, the Soviet demands for German reparations were perfectly understandable from a practical economic point of view because of Moscow's desperate need for capital equipment in the reconstruction and modernization of Soviet industry. But it would be erroneous to ascribe vulgar Marxist ideas to Stalin and believe that he looked at only the economic issues without regard to their political implications. This was acknowledged quite frankly by Vyacheslav Molotov when he said: "The aim of completely disarming Germany militarily and economically should also be served by the reparations plan. The fact that until now no such plan has been drawn up . . . and the fact that the Ruhr has not been placed under inter-Allied control, on which the Soviet government has insisted . . . , is a dangerous thing *from the point of view of safeguarding future peace and the security of nations*." [Emphasis added.][5]

By 1948 several trends had emerged that counteracted the Soviet approach of weakening Germany. Faced with the inability to reach agreement with the USSR on the future of Germany, the three Western powers had focused on the formation of a separate West German state, currency reform, the rebuilding of the West German economy with the help of the Marshall Plan, and, last but not least, potential participation of the country in future West European security arrangements. The dangers inherent in this Western allied course of action were immediately recognized by the Soviet Union. In Soviet eyes the Marshall Plan was "not designed to forestall a possible resurgence of German aggression but to encourage it."[6] The declaration of the Warsaw conference of June 24, 1949, concerning the Western conference stated in summary fashion that "the London decisions are designed not to avert a repetition of German aggression, but to transform the Western part of Germany, and particularly the heavy industry of the Ruhr, into an instrument for the rebuilding of Germany's war potential, to be used for furthering the strategic aims of the United States and Great Britain. . . . Such a plan cannot but create favorable conditions for the recrudescence of German aggression."[7]

It could be argued that such statements are merely propagandistic and do not express genuine fears and anxieties of the Soviet leadership. It would be difficult to make a good case for such an argument. Stalin has been reported to have expressed considerable pessimism about the likely efficacy of political and economic measures to curb the military and industrial potential of Germany.[8] From this perspective, policies aimed at weakening a unified or divided Germany could appear only as an ineffective holding operation.

Division and Dismemberment

In light of all the facts available, it would be erroneous to say that the Soviet leadership was consciously and consistently striving to dismember or divide Germany. On the contrary, as the end of the war approached, Stalin had increasingly rejected this course of action, no doubt realizing that wartime coalitions in Europe historically have tended to disintegrate rapidly after victory and that interallied agreements would therefore be difficult to enforce. The experience of Versailles had clearly shown that controls over Germany were ineffective. In fact, the very issue of interallied control had been a powerful stimulus to revisionist and nationalist tendencies, even though the territorial reductions of Germany had then been quite limited.

Soviet leaders recognized the resilience of German nationalism and the difficulty of enforcing a division of Germany. The view that Stalin did not intend to divide Germany is supported by the extent of German territory transferred to Poland and the scale of the German population expelled. If Soviet postwar policy had provided for a sovereign and viable East German state under Soviet tutelage, it would have been better to form an entity of roughly the same size and power potential of Poland rather than limiting it in territory, population, and material resources to an extent that seemed to exclude the possibility of successful development and long-term viability.

Thus, from the Soviet perspective in the postwar period, considerable problems would be posed by division and dismemberment of Germany. It probably appeared preferable to be included in decision-making concerning Germany as a whole rather than to be excluded and faced with a hostile West German state.

Neutralization of Germany

It would seem, therefore, that analysis of the pros and cons of the German problem advised a policy aimed at maintaining a unified German state—neutralized and noncommunist in character but ranging somewhere between socialism and capitalism in its structure, with a small army and police force for internal security and self-defense. This is the kind of policy essentially suggested by Stalin in his note to the three Western powers on March 10, 1952, in proposals made by his successors in 1954, and, as applied to Austria, in the State Treaty of 1955. This option probably was not foreclosed in the late 1940s.

However, successfully carrying out such a course of action required repudiation of the "two-camp" theory with all its implications of militant tactics. Equally important, it presupposed trust in the peaceful intentions of the Western powers. It also made a necessity of confidence in the peaceful development of a united Germany. Above all, a nonrevisionist united Germany seemed to precondition the return of the areas east of the Oder and Neisse rivers, areas that many Germans—with Western allied encouragement—had already begun to consider as only temporarily under Soviet and Polish administration. Apart from this territorial issue, three years of Soviet occupation had done much to damage the possibility of reconciliation between Germans and Russians. In light of such circumstances, it is not surprising that Stalin rejected this option.

Given these apparently insoluble dilemmas for Soviet security in the postwar era, it would have been entirely understandable if Stalin had drifted into acceptance of the division of Germany as inevitable, trying to contain emerging dangers by pursuing conciliatory policies. Instead, by imposing a blockade on Berlin in 1948, he adopted a strategy of coercion, using local conventional superiority in Europe and attempting to gain Western compliance to ill-defined Soviet demands. It is not perhaps possible for this strategy to be explained entirely in rational terms. Beyond the realization that Germany had been a threat to Soviet security in the past and that it was likely to be one in the future, Stalin may never really have had a clear vision about how to approach, let alone solve, the German security problem. This would explain much of the failed Berlin venture.

The 1950s and Early 1960s: Reemergence of the German Threat?

The Stalin Legacy

The consequences of this failed foreign policy venture, as well as that of another—the Korean War of 1950–52—were severe. Many Soviet anxieties about security vis-à-vis Germany, which in the early postwar period had probably been viewed as a medium- or long-term concern, quickly seemed to become reality. In particular, in the 1950s the Soviet Union was faced with a disadvantageous correlation of military forces, one that was apparently shifting in an even more unfavorable direction.

Even though the Soviet Union had succeeded in exploding a nuclear device in August 1949 and a hydrogen bomb four years later, it lacked the kind of intercontinental delivery system that would effectively put the United States at risk. In essence, the adversary superpower remained invulnerable. The United States enjoyed unchallenged naval supremacy in the Atlantic and Pacific oceans. It also possessed vastly superior scientific-technological and economic resources. In fact, until the late-1950s it remained the *only* superpower, both economically and militarily.

From Moscow's perspective, the most dangerous development in the 1950s was the fact that this power was now, after the double impact of the Berlin crisis and the Korean War, committing itself to a large security role in Europe. The blockade of Berlin was not yet over when the North Atlantic Treaty was signed on April 4, 1949. Furthermore, rather than completely dismantling its military forces and "bringing the boys home," the United States was reintroducing large forces on the European continent. At the outbreak of the Berlin crisis in June 1948, U.S. armed forces in Europe consisted of about 108,000 officers and men; by 1953 the number had reached 427,000.

The obvious concern from the Soviet perspective was the possibility that U.S. economic and technological resources, maritime supremacy, nuclear weapons, and conventional forces in Europe would be combined with West German economic and military potential; that "U.S. imperialism" would use West Germany as a springboard for aggression against the Soviet Union; or that, conversely, a rearmed West Germany would be intent on taking advantage and regaining its lost territories in the east and would draw the United States into war.

The implications of U.S. ability to operate from bases abroad, notably in Europe, were detailed by the Soviet embassy in Washington in reference to a speech by General Brian Kenney (commander of the Strategic Air Command) and an article published about the speech in *Newsweek*. The embassy charged that the article (and Kenney's speech) had

> set forth a plan to use American air forces, air bases and atomic bombs against the Soviet Union, particularly for the destruction of Soviet cities such as Moscow, Leningrad, Kiev, Kharkov, Odessa, and others.... It is further stated in the article that American strategists are thinking in terms of "closing the circle of air bases around Russia" in order to "make it smaller and smaller, tighter and tighter, until the Russians

are throttled." This plan envisages combined air, naval and ground operations from American bases located near the Russian mainland and their use for intensive bombing raids and attacks by guided missiles.[9]

In light of such anxieties and the clear recognition in Moscow that numerous plans existed for the rearmament of West Germany and its inclusion in a European Defense Community (EDC), initiatives were called for to prevent, or at the very least delay, such a development. Stalin's note of March 10, 1952, could be interpreted as such an initiative.

Initiatives to Forestall the Rearmament of Germany

The note dropped the previous Soviet insistence on a completely disarmed Germany and raised the prospect of unification in return for neutralization and the liquidation of foreign military bases on German territory. Such a unified Germany would not be allowed to enter any coalition or military alliance directed against the Soviet Union, it would be prohibited from harboring "organizations hostile to democracy and the cause of maintaining peace" and the size and weaponry of its armed forces as well as its arms production would be strictly limited.

In retrospect, bearing in mind the experience of Austria and Finland, it would seem that Western assumptions about the future orientation of a unified Germany were erroneous. But conversely, it remains doubtful whether Stalin in 1952 had overcome the schizophrenic attitude that had characterized his stance in the 1940s, notably in the negotiations with the three Western ambassadors in Moscow in August 1948. Then, as probably in 1952, he wanted to prevent the rise of a West German militarized and revanchist state. But at the same time he was reluctant to relinquish those German territories that he held firmly in his hand and consent to the establishment of a unified, neutralized Germany that might gradually drift into the Western camp. Thus, even if the 1952 Soviet offer was genuine and the Western powers had tested it, the result of the negotiating process could have turned out to be no different from that obtaining in the 1940s, namely, failure to reach agreement on the German problem.

The same ambiguity applies to the Beria affair immediately after Stalin's death in 1953 and the "new course" adopted by Georgii Malenkov. On the basis of the available evidence, it seems that neither Lavrenti Beria in 1953[10] nor Malenkov in 1954 were any more prepared than Stalin to give up East Germany. As for the proposals put forward by the collective leadership under Malenkov in 1954, they—like Stalin's note—were timed to prevent or delay the entry of the Federal Republic of Germany (FRG) in NATO.

Nikita Khrushchev did not find himself faced with similarly difficult choices as had his predecessors. West Germany was being firmly integrated into the Western military alliance (NATO) and the European Economic Community (EEC). East Germany was made a member of the Warsaw Pact and the Council for Mutual Economic Cooperation (CMEA), and since then the Soviet leadership has firmly committed itself to the survival and viability of the GDR. The threat to Soviet security potentially emanating from West Germany and its participation in the new Western Alliance after May 1955 had to be managed—somehow—within the framework of the two German states.[11]

Hungary and Berlin

Acute and serious threats to the postwar gains made by the USSR in Europe were soon to develop. From the Soviet perspective, they probably derived from (1) the buildup of U.S. conventional and nuclear power in Europe; (2) the growing strength of the West German *Bundeswehr*; (3) the propagation, by U.S. Secretary of State John Foster Dulles, of the "rollback" of communism and, by FRG Chancellor Konrad Adenauer, of a "policy of strength," both of which terms seemed to imply the use of force to achieve reunification of Germany on Western terms; and (4) growing instability in Eastern Europe as a result of a severe erosion of the authority of the Communist parties after Stalin's death. These threats became clearly visible in Poland and Hungary in 1956. If there was ever any time when the Soviet Union was most vulnerable, it was that year. However, the moment of greatest danger safely passed. As became evident, the rollback was only rhetoric and after the Hungarian revolution the rhetoric was abandoned so as not (again) to give rise to illusions in Eastern Europe.

A second moment of danger came in 1961. On the basis of perceived changes in the correlation of forces in favor of socialism, Khrushchev had set an ultimatum in 1958 for the conclusion of a German peace treaty. Its main purpose obviously was to provide for stabilization of the GDR and liquidation of the Western military presence in West Berlin. By 1961, however, developments in the GDR were rapidly drifting toward collapse. As a result of ideological orthodoxy, political repression, and economic hardship in the GDR and fear that the open borders in Berlin would be closed, the westward flow of people became a torrent. Khrushchev felt constrained to act quickly and authorized the GDR to build the wall.

According to Colonel Oleg Penkovsky, the Soviet military was genuinely worried about the big risk Khrushchev seemed to be taking in Berlin,[12] and the party leader himself was concerned and confessed "We had our doubts about the ability of the [East] Germans to control their own borders. The guards were equipped with firearms, but it's not so easy for a soldier to shoot a fellow German. We expressed this concern to our German comrades."[13] In the event, Soviet anxieties were alleviated. The East German border troops and workers' task forces remained disciplined and loyal to the communist regime. Similarly, the Western allies as well as the political leaders in West Berlin (such as Mayor Willy Brandt) and West Germany (including Chancellor Konrad Adenauer) urged restraint and the taking of political and economic measures rather than military moves in response to the building of the wall. Thus, the road was opened for greater consolidation of the status quo in central Europe and a further lowering of Soviet threat perceptions.

Successful Containment of the German Problem: From Khrushchev to Brezhnev

Soviet Military Power and Its Role in Central Europe

Soviet analysts believed the Berlin crisis had two important consequences. First, "the rug was pulled from under the feet of the adventurist elements who had hoped to kindle a military conflict at the open border between the GDR and West Berlin"; in West Germany, Adenauer's approach to the Soviet Union based on "policies from positions of strength" increasingly came to be seen as unworkable.[14] Second, the building of the wall "significantly consolidated the domestic situation in the GDR and contributed to the successful building of socialism in that country."[15]

But why did Stalin's and Khrushchev's fears about the dangers attached to a rearmed West Germany prove unwarranted? In Khrushchev's—and most likely in Leonid Brezhnev's—view it lay in the growth of Soviet military power. Briefly to trace its evolution, even in the early postwar period the Soviet Union had been able to use one major military asset to safeguard its own security, retain control over Eastern Europe, limit the diplomatic advantages inherent in American strategic nuclear superiority, and influence political developments in Western Europe: Soviet preponderance in conventional power. As most observers realized, the asymmetries in Soviet and Western conventional power were such that the USSR would have been able to overrun Western Europe. To that extent, politically Western Europe was taken hostage and was to be induced to help restrain the United States vis-à-vis the Soviet Union.

This hostage function of Western Europe was enhanced by the Soviet Union's transformation into a formidable nuclear power. At the theater level, this transformation began between 1957 and the early 1960s when Soviet forces at tactical air army level were being equipped with nuclear weapons and when guided missiles (Scud) were being deployed at the front and army levels, and artillery rockets (Frog) at the division level. For its INF systems, after 1950 the USSR built up a large force of medium-range (rather than long-range) bombers. Similarly, because of technical difficulties in its early intercontinental ballistic missile (ICBM) program, as well as a possible underestimation of the U.S. buildup in the early 1960s, the Soviet leadership decided to develop a large SS-4 and SS-5 medium-range missile force. At the same time, claims of intercontinental nuclear power served to induce a new sense of vulnerability in the United States and to increase anxieties in Europe.

The alleviation of security concerns as a result of the transformation of the Soviet Union into a major nuclear power was clearly acknowledged by Khrushchev. As he was to say later, "No longer were we contaminated by Stalin's fear.... Now it was our enemies who trembled in their boots. Thanks to our missiles, we could deliver a nuclear bomb to a target at any place in the world. No longer was the industrial heartland of the United States invulnerable to our counterattack. We wanted to exert pressure on American militarists—and also influence the minds of more reasonable politicians."[16] Pressure was to be exerted, of course, not only on American "militarists" but also, and perhaps even more so, on Europe policymakers and public opinion. This design to safeguard Soviet security and enhance Soviet influence was first used during the Suez crisis in 1956, when Khrushchev issued nuclear threats against Great Britain and France. It was also applied during the protracted controversy over Berlin between 1958 and 1961, when he threatened that in the event of war, NATO military bases in various European countries would be destroyed by Soviet nuclear strikes;[17] Germany, he warned, would be "reduced to dust."[18]

In retrospect, Khrushchev held such threats to have been effective: "'If a third world war is unleashed,' Adenauer often said, 'West Germany will be the first country to perish.' I was pleased to hear this, and Adenauer was absolutely right in what he said. For him to be making public statements was a great achievement on our part. Not only were we keeping our number one enemy in line, but Adenauer was helping to keep our other enemies in line, too."[19]

In Khrushchev's perspective, gains had been made in Berlin. The West had "swallowed one bitter pill."[20] Provided the United States could be pressured more directly, confronted with a more credible threat, and made to feel more vulnerable, conditions in central Europe would perhaps become "more mature." The West might then be prepared to swallow yet another "bitter pill." Undoubtedly, this was part of the reasoning underlying Khrushchev's attempt to deploy INF missiles in Cuba.

The lesson that Brezhnev and his colleagues drew from this failed venture, however, was *not* that military power in the nuclear age was ineffective. On the contrary, military power came to be regarded as one of the main tools with which to advance the claim to political equality with the United States and to play a larger role in global politics. In Eastern Europe, as demonstrated by the Warsaw Pact intervention in Czechoslovakia in 1968 and the "Brezhnev doctrine" enunciated in its wake, Soviet military power remained necessary for enforcing discipline. Toward the West European countries—above all West Germany—it served as an instrument with which to try to gain acceptance of the status quo in Eastern Europe, to establish a code of conduct in the relations with the "socialist community" (notably noninterference in internal affairs), and to influence their domestic and foreign policies in directions favorable to the Soviet Union.

Most important, the evolution of detente in the late 1960s and early 1970s was authoritatively explained, and probably widely believed, in the Soviet Union as being the result of significant "changes in the correlation of world forces,"[21] meaning primarily a shift in the military balance in favor of the Soviet Union. As G. A. Arbatov, one of the main theoreticians of East-West detente, claimed, if many of the imperialist powers were now becoming partners in efforts to lessen the threat of war and to normalize relations with the socialist countries, this was "not because of any change in the class nature of their policy."[22] It was because of the fact that these powers have had "to adapt their internal and foreign policies to objective realities," that is, "to changes in the correlation of forces in favor of socialism."[23]

As for the alleged specific effects of changes in the strategic balance between the superpowers on Europe, Soviet analysts were to write as late as the mid-1980s that

> in conjunction with the liquidation of the strategic invulnerability of the United States, the belief of the West European countries in the so-called "nuclear guarantees" of its transoceanic partner was being eroded. Europe began to recognize what a [big] catastrophe would be a contemporary nuclear-missile war for the continent. From this stems the general interest of the Europeans in the avoidance of a military conflict, in the abstention from military-political confrontation, and the development of diverse contacts between Eastern and Western Europe.[24]

Before the various elements of Soviet policy toward West Germany and Western Europe could be integrated into an overall concept of collective security in Europe and could lead to an all-European security conference, however, the Brezhnev leadership considered it appropriate to increase the pressure on Bonn to modify its *Ostpolitik*.

Pressures on West Germany

Some modifications were implemented by the grand coalition government of Kurt-Georg Kiesinger's Christian Democratic Union (CDU) and Willy

Brandt's SPD in the 1966–1969 period. The modifications included the following elements:

- The willingness of the federal government to enter into negotiations with all the European communist states for a normalization of relations, including the establishment of diplomatic relations.
- The consent to establish contacts with the GDR at all governmental and nongovernmental levels.
- The inclusion of the GDR in an agreement on the renunciation of force without, however, extending international legal recognition to the GDR.
- Abandonment of claims that Germany continued to exist as a "legal entity within the borders of 1937" with the FRG as the sole legal successor of the former German *Reich*.
- Adoption of the position that the 1938 Munich agreement was invalid ex post facto and that it was concluded by the threat of force.

These modifications in the West German *Ostpolitik*, however, did not suffice in the view of the Brezhnev leadership. The Soviet Union apparently calculated that if it were to exert pressure on the FRG, more concessions would be forthcoming; the status quo in Europe would become more deeply ingrained in the German national consciousness.

On the basis of such perceptions, the Soviet leaders attempted to isolate West Germany, both in its relations with the East and within the Western Alliance. This Soviet campaign had several facets. First, French President Charles de Gaulle's initiatives within NATO, his exit from the Alliance's military organization, and the ensuing Franco-Soviet rapprochement were extolled; West Germany was told to follow the French lead.

Second, the Soviet leaders construed an alleged "U.S.-FRG axis" as a major threat to European security and world peace by declaring that each partner "conspires to use the other for its own goals." The United States was using the FRG and the German problem" as a "pretext with which to continue the stationing of troops in Europe and as a lever with which to influence directly politics and economics in Western Europe, the Federal Republic is using the United States for the realization of its revanchist plans to change the map of Europe."[25]

Third, the Soviet leaders appealed to latent anti-German anxieties in both Eastern and Western Europe. Brezhnev, for instance, reminded the "peoples of Europe" that "the threat stemming from the aspirations of the West German revenge seekers" was being carefully watched because "in the course of half a century many European countries have twice been the victims of German aggression."[26]

Fourth, the Soviet leaders refused—officially at least—to differentiate among the two major political parties in the FRG. At the Conference of the European Communist and Workers' Parties in April 1967, Brezhnev had advocated broadening cooperation between communists and social democrats. The SPD, however, was in practice excluded as a possible partner. The main reason ostensibly was that the party had fallen into the hands of "rightist leaders" since the promulgation of its 1958 Godesberg party program and was fully integrated into the West German "monopoly-capitalist system."[27] The most detailed study of the SPD published at that time failed to mention that it was precisely the SPD that was willing to develop contacts with communist parties, including the ruling East German Socialist Unity Party (SED).[28]

What little good that was said of the SPD was said also of the liberals, the Free Democratic Party (FDP). Prior to 1969 when the FDP was to form the government together with the SPD and Moscow adopted a different line, a study of the FRG's foreign policy and the role of political parties charged that "the West German monopolies hold the FDP in reserve. It can be brought into action any time if it should turn out that the CDU is incapable of executing the plans of monopoly capital in the FRG."[29] The "progressive forces," in Moscow's perspective, could be found only in the German Peace Union (DFU) and the Communist Party of Germany (KPD). Both parties, however, represented only a small fraction of the total electorate.

Fifth, the Soviet leaders asserted that neo-Nazism was on the rise again and that the West German government had "much in common with the political aims of the neo-Nazis of all shadings."[30] Moscow drew on the fact that from 1966 to 1969, the rightist *Nationaldemokratische Partei Deutschlands* (NPD) was able to poll more than 5 percent of the votes in some of the parliamentary elections in the German *Länder* (states) and thus be represented in the state legislatures.

A sixth argument levelled against the West German government and its evolving new *Ostpolitik* was the assertion that because of its allegedly "militarist," "revanchist," and "neo-Nazi" character, the FRG could "not claim the same equal status" as other sovereign states.[31] The Soviet government gave this argument an ominous turn by demanding, in essence, the right to intervene in German affairs. It did this on the basis of Articles 53 and 107 of the United Nations Charter, which, as *leges speciales* to Article 2 (*lex generalis*), sanction coercive measures against a former enemy state of the anti-Hitler coalition.[32]

A final issue used against the West German government was its alleged quest to gain access to nuclear weapons and, connected with this, its refusal to sign the nuclear nonproliferation treaty. This issue is addressed later in more detail.

What were the results of Soviet pressures on Bonn? By and large, it seems that they were negligible. By 1969 it was evident that the attempt to isolate West Germany within the Western Alliance had failed. On the one hand, de

Gaulle was far from giving the Franco-Soviet *entente* an anti-German twist; on the other hand, the FRG adhered to its close ties with the United States. Due to West Germany's growing economic strength, its political influence in Western Europe was rising. New efforts were made at integration and cooperation in Western Europe and NATO, with the full inclusion and participation of West Germany. At the same time, the electorate was ready for a change of power in Bonn—a fact that had much to do with the internal divisions and weakness of the CDU leadership and very little to do with Soviet policy. As a result, any "scientifically based" approach in Moscow sooner or later would have had to abandon the tactic of circumventing and isolating the Bonn government and refusing to deal with the country's main political forces.

Conditions in Eastern Europe were also ripe for a new Soviet perspective on Germany. From the vantage point of the Brezhnev leadership, the important security dimension of the Prague Spring in 1968 rested in the combination of Czechoslovak ideological reformism and revisionism; the rising attractiveness of West European, notably West German, social democratic ideas; and the danger that a nonconformist Czechoslovakia would yield to the pervasive attractiveness of West Germany and its economic influence and embark on extensive cooperation with the country, perhaps even to the point of drifting away from CMEA and the Warsaw Pact. The intervention of the five Warsaw Pact countries in August 1968 removed such threats to the Soviet Union's position in Eastern Europe. Thus, perhaps paradoxically, it untied the hands of the Soviet leadership and facilitated a more favorable response to the *Ostpolitik* of the new coalition government between the SPD and FDP, formed at the beginning of October 1969.

Detente and Its Demise: Emergence of New Threats in the 1970s and 1980s?

Soviet security perspectives on the SPD/FDP coalition government under Willy Brandt were of lessened dangers. As mayor of West Berlin, Brandt had done much to prevent the tense situation of August 1961 from exploding; above all, he had refrained from inciting the German population on both sides of the sectoral border to commit acts of violence. In his inaugural address, the new chancellor—for the first time in the history of official government statements to the *Bundestag* (the lower house of parliament)—spoke of "two states in Germany," thereby taking another step toward accepting postwar realities. Almost immediately after coming to power, the Brandt government also signed the nonproliferation treaty.

The Secretary General of the Communist Party of the Soviet Union (CPSU) committed himself to new appraisals by declaring that the formation of the coalition government led by the SPD represented a "significant change in the constellation of political forces in the German Federal Republic."[33] The Moscow meeting of the political consultative committee of the Warsaw Pact, held in December 1969, favorably noted trends in West Germany that were "directed at a realistic policy of cooperation and understanding" in Europe and characterized Bonn's signature to the nonproliferation treaty as a "positive element."[34]

Throughout the 1970s, West Germany participated fully in the process of East-West detente; joined the Conference on Security and Cooperation in Europe (CSCE); concluded treaties for the normalization of relations with the Soviet Union, Czechoslovakia, and Poland; set its relationship with the GDR on a new footing; facilitated the Quadripartite Agreement on Berlin; played an active role in arms control negotiations, notably on Mutual and Balanced Force Reductions (MBFR) in Vienna; provided an important impetus to East-West economic cooperation; and set some of the highest growth rates of the Western industrialized countries in trade with the USSR.

It is not surprising, therefore, that during this decade West Germany became an *interlocuteur privilégié* of the Soviet Union. Consequently, Soviet analysts viewed the relations between the socialist countries and the FRG in the 1970s as an "important factor of stability and good-neighborliness in Europe."[35] D. E. Mel'nikov, perhaps the foremost Soviet expert on the German problem, predicted in 1973 that

> for the FRG, the 1970s will be a period of the further deployment of political forces in all directions, a period of sharp struggle between reaction and progress. The outcome of this struggle will also affect the foreign policy line of the Bonn government. There can be no doubt that an important factor in this respect consists in the fact that the *majority of the West German population stands on positions of the recognition of realities, and desires peace and good-neighborly relations with all the peoples of Europe.*[36]

The perception of a conciliatory tendency persisted despite the fact that the Soviet-American relationship, starting from the October 1973 Arab-Israeli war, had begun to deteriorate and by the end of the decade had ended in sharp mutual recrimination. As Soviet analysts have noted, "a marked complication of the international situation occurred."[37] Such complications included, in their view, the discovery of an alleged Soviet combat brigade on Cuba in the fall on 1979; NATO's December 1979 decision to deploy INF missiles in Western Europe; the shelving of the Strategic Arms Limitation Talks (SALT II) treaty ratification and the imposition of sanctions against the Soviet Union under the "pretext" of a response to the dispatch of a "limited contingent" of Soviet forces to Afghanistan in December 1979; and the renewal of sanctions against socialist countries in alleged reaction to the declaration of martial law in Poland in December 1981.

It should be noted that it was primarily the administrations of Jimmy Carter and Ronald Reagan that were held responsible for the deterioration of East-West relations; until May 1984, West Germany was largely exempt from propagandist attacks. The West German coalition government of social democrats and liberals under Chancellor Helmut Schmidt was only found "guilty by association." As the former Soviet ambassador to the FRG, Valentin Falin, speaking about plans for stationing INF missiles in Western Europe, put it: "There are two violins in the [NATO] orchestra. The first violin is played by the Americans, [only] the second by the Federal Republic."[38]

The main thrust of analysis concerning that period, therefore, was to comment favorably on the dominant tendency in West Germany to save as much substance as possible from the preceding period of East-West detente,[39] to try to prevent "horizontal escalation," and to forestall the spilling over of conflicts in the Third World (such as in Afghanistan and the Persian Gulf) into central Europe.[40] Such attitudes of the Schmidt government, as Soviet observers noted, were criticized in Washington as fostering the unacceptable notion of the "divisibility of detente."

Given the dominant line of analysis that West German policies were generally directed at safeguarding security through cooperative efforts with the East, it is (on the surface) perhaps surprising that starting in May 1984, West Germany became again the target of vicious Soviet propagandistic attacks. "Revanchism" and "militarism" were seen to be on the rise again.[41] West Germany came under attack for following the U.S. militarist course, endorsing the strict and rapid implementation of INF deployment, fostering again the notion of the "unsolved German problem," and questioning the political and territorial realities in Europe. It was therefore also held responsible for a deterioration in the relations between the FRG and the socialist countries.[42]

Did the Soviet leadership under Yuri Andropov and Konstantin Chernenko really believe in the reemergence of a "German threat"? Did it, in fact, proceed from the assumption that the risk of war in central Europe had increased because of a convergence of aggressive policies of the U.S. and FRG governments under Ronald Reagan and Helmut Kohl? Why had the previous conciliatory trend in West German politics disappeared, and why so suddenly?

No satisfactory answers are given to such questions, at least not in writing. In conversations with Soviet scholars, however, it is sometimes possible to confirm one's own interpretation. Starting from the second half of the 1970s, under Brezhnev and continuing under Andropov and Chernenko, the military factor in Soviet foreign policy had become ever more important. The three Soviet leaderships apparently thought that forcing the pace of the military competition in Europe and improving the Soviet preponderance in conventional and INF systems would make it possible to translate military power into political influence. A relentless buildup of military strength and political intimidation accordingly was meant to change the domestic political makeup and foreign-policy orientation of the West European countries. The role of nuclear hostage was to be impressed on them, that is, that they exposed themselves to the risk of nuclear retaliation by the USSR if they decided to support the United States in a military conflict.[43] The rise of a powerful peace movement in Western Europe, foremost in West Germany, was probably regarded in Moscow as one of the main indications that this approach was working.

What Soviet leaders did not appreciate, however, was that increased Soviet pressure also produced counteractions. The failure by Brezhnev and Andropov to make substantial concessions on the INF issue significantly contributed, in the fall of 1982, to the demise of the Schmidt government and the SPD, the realignment of the FDP, and—with the help of the FDP—the return to power of the CDU. Perceptive Soviet observers had warned long ago that U.S.-FRG cohesion is threatened when international tensions abate[44] and that "the general law of the development of the mutual relations between the FRG and the USA is that they tend to become the closer and more indissoluble, the stronger the heat of the struggle between the two world systems."[45] Indeed, this is precisely what happened after the INF deployment controversy within NATO had subsided. Furthermore, the peace movement, which had come to be seen in Moscow as an effective instrument for changing government policy in Western Europe, including West Germany, began to decline almost immediately after the stationing of INF missiles had begun.

Declining Effectiveness of the Peace Movement

In 1984 one could still read that the Western peace movement was far from defeated. For instance, two Soviet analysts reported that in June 1984, an opinion survey had taken place in West Germany "with 150,000 activists polling 5 million citizens of which 87 percent opposed the stationing of new intermediate-range nuclear missiles on the territory of the Federal Republic and who supported the withdrawal of those [missiles] already emplaced there."[46] Another Soviet observer acknowledged that the "peace movement had lost the campaign against cruise and Pershing missiles," but, he claimed, it was entering "a new stage of development" and gradually transforming itself into a permanent political factor that would again be able to exert "effective influence" over government decisions.[47]

By the time of the 27th Party Congress such optimism had faded. Gorbachev's report even contained language that had to be interpreted as criticism of the previous approach. His message was: "Ensuring the continuity of its foreign policy strategy, the CPSU will pursue an active international policy which proceeds from the realities of the modern world. Of course, it is *not possible to solve the problem of international security with one or two, even very intense, peace offensives*. Only consistent, systematic and persistent work can bring success." [Emphasis added.][48] Soviet observers subsequently reinforced this analytical trend.[49]

Having reconstructed Soviet perspectives on the more general, historical, and political aspects of German security, what are the prevailing Soviet views on more specific military aspects?

The West German Armed Forces: Strength, Composition, and Doctrine

Strength of the Bundeswehr

Given its location, sizable population, and industrial base, as well as its status as a nonnuclear power, it was almost logical that West Germany would become the mainstay of NATO's conventional defense in central Europe. Soviet analysts note that West Germany provides "half of all the ground forces of the North Atlantic bloc in Central Europe, and also approximately one fourth of the air forces of NATO. In the Baltic, the West German naval forces today furnish 70 percent of the fleet and all of the naval aviation of NATO."[50] No less than 10 percent of the total chemical arsenal of the United States was stationed in West Germany.[51] Owing to an efficient military infrastructure and the availability of reservists, the *Bundeswehr* could in a very short time expand to an army of several million soldiers.[52] Not all Soviet observers, however, are convinced of the efficiency and effectiveness of the military reserve system of the FRG, although the *Bundeswehr* is seen to be trying to overcome these deficiencies.[53]

Training in the Spirit of Revanchism and Militarism

The training and indoctrination in the *Bundeswehr* is allegedly designed to foster hate against the socialist countries—and especially the Soviet Union—as demonstrated by the supposed fact that for "shooting practice the targets represent Soviet soldiers and their combat gear."[54] Education proceeds "in the spirit of revanchism and militarism": "The soldiers are being instilled with the notion that communism by its very nature is 'aggressive,' that it threatens 'Western freedom' and that the FRG has a lofty mission in its defense. A state of affairs is being presented in such a way as if the Soviet Union would at any moment move its troops against West Germany and other NATO countries and as if there were no choice other than to prepare for war against this 'main enemy.'"[55] In hard-line and propagandist Soviet discussion and in the indoctrination of young people (for example, *Komsomol* publications), the impression is being conveyed that there is little difference between the present West German armed forces and Hitler's *Wehrmacht*.

Doctrine

As if to underline a direct parallel with the Nazi past, the claim is being made that the West German *Bundeswehr* has expansionist territorial goals and an offensive military strategy. Semantic tricks are used to give substance to this assertion. This applies, above all, to the German term *Vorneverteidigung*, that is, the notion that defense is to begin *at* the border between West and East Germany, and that as little Warsaw Pact penetration of West German territory as possible should be allowed. This concept is rendered in Russian by *strategiya vydvinutykh rubezhei*[56] or *kontseptsiya peredovykh rubezhei*.[57] This translates as the "concept of forward borders" and is interpreted to mean that war in its initial phases is to be carried, with conventional or nuclear weapons, *beyond* the West German borders to the territory of the Warsaw Pact countries. It also conveys the notion that the borders already have been or will be advanced or pushed forward. According to one Soviet definition:

> At the basis of the preparation of the *Bundeswehr* for war lie the aggressive coalition strategy of "flexible response" and, adopted by NATO in 1963 (at the insistence of the military-political leadership of the FRG), the concept of "forward borders" (*peredovykh rubezhei*). According to the strategy of "flexible response" the USSR and the member-countries of the Warsaw Treaty are the main adversary. The concept of the "forward borders" requires the deployment of NATO forces close to the Eastern borders of the FRG and envisages, in the event of a military conflict in Europe, the unfolding of military operations at the forward borders along the frontiers of the GDR and their fast transfer to the territory of the GDR and the CSSR [Czechoslovakia].[58]

Only occasionally does one find the correct rending of the German term as *peredovaya oborona*, but even this is qualified by quotation marks or the addition of "so-called."[59]

In the 1980s, Soviet writers linked the threats construed by misrepresentation of *Vorneverteidigung* to the Rogers Plan, that is to the concept of air-land battle and deep-echeloned strikes. In their view:

> The concept of the "air-land operation [battle]" began to be intensively developed in the U.S. armed forces as far back as the early 1970s. To a considerable extent these developments were oriented toward reviving the "offensive spirit" as a result of the prolonged, and lost, war in Vietnam. The development of a new concept for the U.S. ground forces was regarded as a means of overcoming the "excessively passive" nature of the concept of "forward defense," which is part of the NATO strategy of "flexible response."[60]

In this way, West Germany's—and NATO's—aim of stopping the first echelon of Warsaw Pact forces and attempting to disrupt the advance of the second and third echelons is completely turned on its head. In another analysis, the Rogers Plan is regarded as aiming at NATO's achievement of superiority over the Warsaw Pact in the conventional sphere. This plan is said

> to foresee a significant build-up of conventional weapons in the countries of the North Atlantic bloc, mass deliveries of highly precise weapons to the troops, and their [the weapons'] broad use in operations in the European TVD [theater of military operations] with the aim of inflicting deep strikes on the adversary and, above all, to destroy its second echelons and its reserves, and preventing the strengthening of the forward groupings. Having declared war, the NATO leadership envisages carrying over the military operations, in the shortest period of time, to the territory of the countries of the Warsaw Treaty.[61]

Soviet military analysts have related the Rogers Plan and the forward and follow-on forces attack (FOFA) concept also to the "accumulation of the nuclear potential of the USA and NATO in Europe," including the deployment of INF missiles in that region,[62] and have charged that "the creation and the deployment of the 'Euromissiles' on the continent is usually connected by foreign military specialists with the concept of the 'forward borders' (*vydvinutykh vpered rubezhei*), which envisages the use of nuclear weapons from the very start of a military conflict."[63] In turn, the use of nuclear weapons, and notably the deployment of INF, was to provide NATO with the capability of inflicting a first strike or disarming strike against the Soviet Union and its allies.[64]

Finally, Soviet writers have linked these concepts to the idea of "horizontal escalation." In the view of one of the more sophisticated civilian analysts:

> The danger of outbreak of a war in Europe as a result of the adoption of the concept of "deep echeloned strike" and "air-land battle" is further intensified in connection with the 1982 adoption by the U.S. military-political leadership of the principle of "horizontal escalation." According to this principle, in the event of a crisis situation in any region of the world in which the United States finds itself in a situation in

which it is not able to gain the upper hand over the forces that are supported by the socialist community, the American state leadership retains for itself the "right" *to unleash military operations in Eastern Europe, that is, against the Warsaw Pact member-countries.* [Emphasis added.][65]

Quest for the Acquisition of Nuclear Weapons?

An objective analysis of Soviet security would probably conclude that the main danger facing the USSR does not lie in a large-scale conventional attack and occupation of the country by NATO forces but in a nuclear strike; that this is, in fact, the *only* threat to the existence of the Soviet Union. It is in all likelihood for that reason that Soviet political leaders and propagandists, as well as military and international relations experts, have regarded any possibility of West German access to nuclear weapons—either in the form of direct possession or participation in decision-making—as particularly dangerous.

Charges to this effect leveled against various West German governments, political parties, and leaders perhaps cannot often be taken at face value. However, they do most likely serve the important purpose of reiterating the Soviet leadership's fundamental objection to German nuclear armament, either unilaterally, in cooperation with France and Great Britain, or in the NATO framework.

One Soviet account typical of such preventive analysis asserts that the West German government had "tried openly, up to the end of the 1960s, to liquidate in one way or another the non-nuclear status of the country."[66] An earlier analysis argues that by the end of that decade, "as a result of the systematic support that was rendered by other imperialist states, foremost the USA, the technical and economic basis required for the production of its own nuclear weapon ... [was] created in West Germany."[67] The author, of course, recognizes that for West Germany the creation of its own nuclear weapon would be "a difficult, expensive and long process." It would also mean "a simultaneous renunciation of the obligations of the Paris agreement, which prohibited West Germany to produce nuclear weapons on its territory." Therefore, he claims, parallel to the development of its own nuclear industry, Bonn strongly seeks nuclear weapons directly from the United States or indirectly through NATO and actually wants to gain control over the U.S. warheads on West German territory.[68]

Bonn did sign (and ratify) the nonproliferation treaty. But this did not lay to rest the issue of German access to nuclear weapons. The issue resurfaced in the context of the controversy over the enhanced radiation weapon ("neutron bomb") and especially during the acrimonious East-West argument over the stationing of INF missiles in Western Europe, with Soviet spokespersons again charging that the FRG wanted to become a nuclear power. Thus, General Nikolai Chervov, a member of the Soviet General Staff, asserted that "the possibility of a transfer of technology to the Federal Republic for the production of cruise missiles" existed and constituted by far "the greatest danger in the secret equipment of the Federal Republic with strategic weapons."[69]

Other observers have dwelled on the risk of the FRG's acquiring access to nuclear weapons through the establishment of a joint European nuclear force.[70] They charge that "Bonn long ago sought pledges from Paris concerning the possible use of French nuclear weapons on West German territory. These are the Pluton tactical missile and the Hades missile ... as well as the Mirage and Jaguar aircraft.... According to press reports, the French government has agreed to hold consultations on these questions with the FRG."[71] An important question that arises for some Soviet analysts, therefore, is: "Will not Franco–West German consultations be a prelude to the assertion that 'France's vital interests' also extend to FRG territory? After all, there is only one step between this assertion and the pledge to expand the French 'nuclear umbrella' to the other side of the Rhine."[72]

Thus, three aspects of possible Franco-German or, more broadly, West European nuclear cooperation have been discussed in Soviet sources. The first is the extension of deterrence through French nuclear weapons. This could be observed when French Prime Minister Jacques Chirac stated that "in the event of an attack on the FRG, France would immediately come to her help. French military experts have demanded from the authorities that they 'clearly and unambiguously' declare that 'the defense of France begins at the Elbe river' and that French tactical nuclear weapons will be used for the defense of all of Western Europe including the FRG."[73] A second aspect is the proposal by former "French Defense Minister [Charles] Hernu to station in West Germany the French neutron weapon (it is ready and only needs to be serially produced) and to give Bonn access to it." A third is the formation of nuclear forces within the concept of a common West European defense.[74]

But other analysts blur such sharp outlines of nuclear cooperation. They offer a more sober interpretation of the state of affairs. For instance, they note that in 1986, "Paris declared that agreement to hold consultations by no means signified consent to FRG participation in adopting a decision on the use of nuclear weapons. The right to use them, they said, remained as before the exclusive prerogative of France's supreme political and military leadership."[75] To this clarification was added another (as early as in the 1970s), namely that "even if a European political union were to be formed, where the nuclear forces of Britain and France would play an important role in the security of Western Europe, the Federal Republic would still put its trust first and foremost in the U.S. nuclear umbrella."[76] Conversely, not without some justification, its author suspects that any British or French notion of a West European nuclear force has "never meant anything but the possibility of an Anglo-French nuclear duopoly and the attempt to maintain West Germany's status as a non-nuclear military power."[77]

West Germany and Military Cooperation in Western Europe

The Western European Union and Independent European Program Group

Soviet analysts have recognized two approaches through which the European NATO countries have attempted to assert their special interests: (1) the Europeanization of NATO and (2) the transfer of various functions of the Atlantic alliance to an autonomous European decision-making entity. In the Soviet view, both approaches require closer coordination, if not integration, of the West European countries in the politico-military sphere.[78]

In Europe's postwar history two multilateral organs were created precisely for this purpose of regulating military cooperation: the Western European Union (WEU), formed in 1954, which includes all the countries of the European Community (EEC) except Denmark and Ireland, and the Eurogroup, created in 1969, which includes the original nine EEC members minus France and Ireland, plus Greece, Turkey, Norway, and Portugal. From the Soviet perspective, neither organization has been able to make much headway in specialization, standardization, and interoperability of weapons systems or in reducing the tremendous disproportion (10:1) in the arms trade between the United States and Europe.[79]

As for the WEU, in the 1960s and 1970s Soviet experts asserted that the main stumbling block to the fulfillment of its stated objectives—gradual integration into the politico-military sphere, joint arms production, and standardization of weapons systems—was the attempt to keep West Germany's military potential and ambitions in check. For this reason many proposals advocated by Great Britain and other countries to revive the cooperative mechanism of the WEU failed.[80] In reality, one should point out, West Germany's reluctance is based not on the fear of discrimination but on the fact that it prefers West European military coordination and cooperation within a NATO framework rather than outside it. But Soviet analysts are correct to state that the FRG's government "has attempted to revive the activity of the WEU" and has "tried to utilize this organ above all for the achievement of closer military cooperation and coordination in the area of arms production."[81]

When the WEU was founded, severe restrictions on the production of certain types of weapons were imposed on West Germany. Soviet observers exaggerate somewhat when they state the restrictions "in no way stood in the way of the rearmament of West Germany."[82] They have attacked with particular vigor the WEU's decision "to lift, upon the initiative of the FRG, the restrictions imposed in accordance with international agreements concerning the production of strategic bomber aircraft as well as long-range [conventional] missiles."[83] Some analysts linked this step with the INF deployment and increasing revanchism and militarism in the FRG and derive from this another serious threat to the Soviet Union.[84]

By and large the importance and effectiveness of the WEU are seen in a more moderate light. According to one summary written in the 1980s:

> At the creation of the WEU, the leadership of the main West European states hoped to use it as a means for developing cooperation between the member states and for increasing their role in determining NATO policy. According to the [1954] Paris agreement, this Union was to facilitate the unity and gradual integration of the West European states in the military-political sphere, the setting up of joint weapons production and their standardization. However, in reality, according to foreign observers, it turned into a merely consultative organ whose decisions do not have an obligatory character for the member states. Furthermore, to a significant degree *the effectiveness of the WEU activity is lowered due to a constant collision of interests of Great Britain, France and the FRG* waging a fierce struggle for leadership in Western Europe.[85]

All this was despite the fact that the WEU, according to some analysts, was to be given new life as a result of NATO's diminished prestige in the wake of the INF deployment and the problem of arriving at common decisions with such "difficult partners" as Greece and Denmark.[86]

As for the Eurogroup, it is argued that the main impediment to its effectiveness has been the absence of France. In addition (presumably for reasons of cost-effectiveness), there had also been instances when the Eurogroup did not present a European spirit but acted as a "supporter of the American monopolies and Washington," for instance in their purchase of U.S. fighter aircraft and acquisition of the Lance short-range missile.[87]

Nevertheless, some progress in the formulation of joint programs for arms production in Western Europe is seen to have been made, the volume of such production amounting to approximately 10 percent of the total arms manufacture in that region. Further expansion of common European programs could follow, it is thought, which would be due primarily to the Independent European Program Group (IEPG) founded in 1976.[88] This prediction could prove accurate, because the IEPG squares the circle of obtaining French participation and satisfying the West German preference for European military cooperation connected to the NATO framework.

On the basis of these observations, the following conclusions emerge for cooperation among the West European states in the sphere of conventional arms production: Despite "centripetal tendencies," the degree of cooperation achieved so far is marginal but likely to increase. There will be an increase in cooperation, but this will not be dramatic, for two reasons. First, "there is a constant contradiction between the tendency of each major West European state to strengthen its autonomy and the scale of its own military industry, and the impulses towards integration.... It appears that in the conditions of the 'Europe of the Fatherlands,' the production of weapons in the future, as today, will be realized primarily on a national basis." Second, "what persists [also] is the system of military-industrial relationships within the NATO framework, where the U.S. plays the main role."[89]

The European Community

From the Soviet perspective, many West Germans belong to those "advocates of military-political integration of the West European countries who advance far-reaching plans for the creation of a new military-political grouping on the basis of the European Economic Community (or 'Common Market'). These proponents proceeded from the fact that such integration can give an impulse not only to economic cooperation" but also to the "transformation of the EEC into a military-political bloc: the European Union."[90]

Along these lines, the prime minister of Belgium, L. Tindemans, had recommended in a report adopted in 1976 giving the EEC military functions. However, Soviet analysts have recognized that not much has come of these plans. Thus, an article written two years after the appearance of the report neither mentioned it nor the prospects of military cooperation within the framework of the EEC.[91] In the 1980s, Bonn was still seen as being "especially disappointed at the postponement yet again of the discussion of the project to establish the European Political Union within the framework of the EEC, the essence of which consists in the Community starting to carry out a common policy on questions of defense and security."[92]

Some analysts, however, do emphasize the potential for closer military cooperation and integration up to and including in nuclear matters. They claim that

> Western specialists do not exclude the possibility of the formation at the EEC of a special organ at the level of the defense ministers. It is thought that the next step in the establishment of the European Union could be the creation of European nuclear forces consisting of the strategic forces of France and Great Britain with financial and technological participation of the FRG, the adoption of a single military

budget, the creation of a unified European command and unified armed forces.[93]

Franco-German Military Cooperation

Soviet analysts note that one factor that has contributed to the revival of the WEU's activities has been the increase in Franco-German military cooperation. Of particular importance has been the strengthening of cooperation in the military-industrial sphere. Thus, for instance, agreement has been reached on joint participation in formulating and implementing a number of space projects, including creation of the Ariane rocket and future generation missile engines designed to launch space stations, as well as military reconnaissance satellites. Agreement has also been reached on joint production of a combat helicopter.[94]

The year 1986 was marked by a further acceleration of military-political cooperation between Paris and Bonn. This too has been duly registered in Moscow. While noting their plans for joint military maneuvers and increased joint arms production, one Soviet analyst indicated that

> the center of gravity ... is being increasingly shifted to coordinating military policy and converging military doctrine. At the end of last year, the French president and FRG chancellor agreed to "unfreeze" the articles of the 1963 Elysee Treaty that concern defense and security problems. This subject matter had been "prohibited" for almost a quarter of a century. Now, within the framework of a bilateral commission, problems of "strategic coordination" are being actively discussed by politicians and military men.[95]

It is difficult to say to what extent Franco-German military cooperation is interpreted by Soviet leaders as affecting the East-West military balance in Europe. Most likely its function is seen to be primarily political, as well as having psychological and symbolic significance. But to the Soviets symbolism *does* matter. They have therefore consistently rejected such cooperation. More recently, their propagandists have charged that Franco-German military cooperation, especially in the nuclear sphere,[96] is meant "to counteract the Soviet peace initiatives, which are directed at the liquidation of nuclear weapons, and to fill the vacuum which could develop after the elimination of American and Soviet missiles in Europe."[97] In their view, Franco-German military cooperation, as with many other schemes of "Euro-defense," is in contradiction to the new thinking required in contemporary international affairs. And to them, "It is quite obvious that it runs counter to the interest of a strengthening of the security of all the states on the continent and has nothing in common with the deepening of confidence and the spirit of our times."[98]

"New Political Thinking" and the German Problem

Only recently have lowered threat perceptions and the acknowledgment of legitimate security interests of the West European countries, including West Germany, found their way into comments and interviews by Soviet officials and writers. This phenomenon is clearly an integral part of the "new political thinking" that Gorbachev has demanded of all the actors on the nuclear stage.

An example of revised Soviet security perspectives has been provided by the former Soviet ambassador to West Germany, Valentin Falin. In talks in Moscow with Willy Wimmer, the CDU spokesman in the West German *Bundestag*, he agreed that the Soviet Union in the past had not adequately assessed the objective state of European security and had failed to respond properly to the subjective security needs of the West Europeans.[99] In an interview with *Der Stern*, a West German magazine, he also declared that "we have lost much time on both sides by the construction of a wrong image of the enemy (*Feindbild*).[100] The current ambassador, Yulii Kvitsinskii, rather than simply demanding West German adherence to the territorial status quo and respect for the security interests of the socialist countries, asserted that the Soviet Union was interested also in the security of the FRG.[101] One Central Committee expert on the German problem, Nikolai Portugalov, conceded that the Soviet Union knew about the security concerns prevalent in the West, notably about the military "muscle power" of the USSR. Such concerns were to some extent understandable and probably in part due "to our overreaction in security matters to the challenges of the West during the years of the cold war and the arms race but also the effects of stereotypes of propagandist origin."[102]

The theoreticians of new political thinking have taken up this theme and pointed out that in the past, notably in the "period of stagnation" (that is, in the Brezhnev era), a number of factors had contributed to the construction of certain stereotypes of the enemy on both sides, East and West. Among the factors mentioned are, above all, "the consideration of the differences and contradictions between the two social systems, and therefore between the [individual] countries [belonging to them], as an absolute given"; adherence to "ideological remnants of the theory of 'world revolution'"; and the "clinging to secretiveness and suspicion, and impenetrable 'monolithism.'"[103] They also ask the pertinent question of whether "our orthodox social scientists, armed with quotations, have not, in extreme moralistic colors, painted the world as an arena in which the 'good' and the 'evil' are struggling with each other?"[104] In their view, it was necessary to rid oneself of the idea that a competitor invariably had to be considered an enemy. One had thus to embark on a "radical departure from the traditions of the past," on a "de-escalation of political rhetoric" and the "emancipation from those of its forms which are most strongly ideological and portraying matters as absolute."[105]

Also instructive for such changes in interpretation, perception, and mood is an interchange between Senator Joseph R. Biden, Jr. (D, Delaware), and the chairman of the Supreme Soviet of the USSR, Andrei Gromyko. The senator was reported by *Pravda* as having said to his Soviet host that "certain political practitioners in the West are of the opinion that one could not safeguard peace without the nuclear weapon. Testimony to this, it is alleged, were centuries in the history of Europe, which constantly fought without the nuclear weapon. Even *Germany*, if it were to be without it, *could quickly try to secure it* [the nuclear weapon] for itself and *again threaten other countries*."[106] Interestingly, Gromyko, who had played a central role in the formulation of Brezhnev's and Andropov's policies toward Western Europe, failed to use the opportunity to emphasize Soviet threat perceptions of West Germany. He pointed out instead that Biden had talked "as if there were only one Germany. But there are two of them: the German Democratic Republic and the FRG. We are convinced that the Germans in Europe cannot consider themselves to be some island isolated from everyone. These two Germanies will be part of a non-nuclear Europe."[107] If anything, such assertions reflect a high degree of confidence, perhaps even arrogance, vis-à-vis East and West Germany.

Some Soviet scholars share this new sense of confidence and lessened threat perception, and they provide the theoretical underpinning for it. In an important article on new and old challenges to security, three academic specialists proceed from the premise that human thinking had a tendency "to lag behind rapidly changing political reality." This explained the "natural tendency of ordinary thought to focus the main attention on traditional threats, which one already has had to contend with, and to underestimate new, previously unknown ones, which have, however, become, or are rapidly becoming, reality."[108] But, they continue,

> when making an assessment of the real military-political situation in the world, it is necessary to avoid considering the past experience as an absolute given, and fully to take into account the changes which have taken place. [This also applies] to the main threat in the mass consciousness of the Soviet people which traditionally has been a repetition, in one form or another, of June 22, 1941, in other words, of an invasion from the West, either a nuclear attack or an aggression with conventional forces.[109]

However, they argue, the nuclear age has given this problem a special dimension. The threat of nuclear annihilation undoubtedly was analogous to 1941, and for that reason one could to some extent legitimately compare the stationing of the Pershing II missiles in Western Europe—capable of reaching Moscow in ten minutes—with the fall of 1941 when Nazi troops had reached the outskirts of the Soviet capital. But there is a crucial difference between the early 1940s and the current situation: "Nuclear attack would in the final analysis mean inescapable catastrophe also for the aggressor." Equally important, "There is today, in the relations between East and West, not a single conflict which, in the quest to solve it, would give rise to the temptation of resorting to war. On grounds of common sense, it is difficult to imagine in the name of what goals Western armies would invade the territory of the socialist countries."[110]

Probably mirror-imaging Soviet dilemmas, Zhurkin et al. point out that contemporary capitalism had to cope with considerable problems, but these problems "in principle cannot be solved by military aggression against socialism."[111] Such views have not yet been reflected in a change of Soviet policy toward Western Europe; there is still a wide gap between discussion and the practical steps taken in force development. But there is a possibility that the new inflections on the German threat and European security will affect and alter the framework in which decisions are taken and will ultimately influence policy itself.

Notes

1 Research for this chapter was completed in May 1988.
2 As quoted by Milovan Djilas, *Conversations with Stalin*, translated by Michael B. Petrovich (New York: Harcourt Brace & World, 1962), p. 114.
3 Ibid., p. 79.
4 Recollection by Charles Bohlen; personal communication to the author.
5 At the July 10, 1946, meeting of the Council of Foreign Ministers in Paris; see V. M. Molotov, *Problems of Foreign Policy: Speeches and Statements, April 1945–November 1948* (Moscow: Foreign Languages Publishing House, 1949), p. 66.
6 M. Marinin in a two-part series on the Marshall Plan, *Pravda*, March 5 and 6, 1948.
7 *Pravda*, June 25, 1948.
8 As quoted by Djilas, *Conversations with Stalin*, pp. 114, 121. Similar views were expressed by a Soviet foreign ministry official to Rudolf Nadolny, a former German ambassador in Moscow, in 1946; see Nadolny, *Mein Beitrag* (Wiesbaden: Limes, 1955), pp. 178–179.
9 As quoted in Department of State, *Foreign Relations*, Vol. IV (Washington, DC: Department of States, 1948), p. 887.
10 See Victor Baras, "Beria's Fall and Ulbricht's Survival," *Soviet Studies*, 7 (1975), pp. 381–395.
11 See P. A. Nikolaev, *Politika Sovetskogo Soyuza v germanskom voprose, 1945–1964 [The Policy of the Soviet Union on the German Question, 1945–1964]* (Moscow: Nauka, 1966), pp. 211–218.
12 Oleg Penkovsky, *The Penkovsky Papers*, translated by P. Deriabin, with an introduction and commentary by Frank Gibney and a foreword by Edward Crankshaw (London: Collins, 1965), pp. 131–132, 136–138, 142, 161.

13 *Khrushchev Remembers: The Last Testament*, translated and edited by Strobe Talbott, with a foreword by Edward Crankshaw and an introduction by Jerrold L. Schecter (London: Deutsch, 1974), p. 508.
14 A. S. Grossman, "The Border of Peace," *Voprosy istorii*, 10 (1969), p. 201; V. G. Trukhanovskii, ed., *Istoriya mezhdunarodnykh otnoshenii i vneshnei politiki SSSR, Vol. III: 1945–1963 gg. [The History of International Relations and the USSR's Foreign Policy, Vol. III: 1945–1963]* (Moscow: Mezhdunarodnye otnosheniya, 1964), pp. 211–212; N. N. Inozemtsev, *Mezhdunarodnye otnosheniya posle vtoroi mirovoi voiny, Vol. III: 1956–1964 gg. [International Relations after World War II, Vol. III: 1956–1964]* (Moscow: Politizdat, 1962), pp. 523–534.
15 Grossman, *Voprosy istorii*, 10 (1969), p. 201. Similarly, G. M. Akopov, *Zapadnyi Berlin: problemy i resheniya [West Berlin: Problems and Solutions]* (Moscow: Mezhdunarodnye otnosheniya, 1974), pp. 164–255; P. A. Abrasimov, *Zapadnyi Berlin: vchera i segodnya [West Berlin: Yesterday and Today]* (Moscow: Mezhdunarodnye otnosheniya, 1980), p. 46; V. N. Vysotskii, *Zapadnyi Berlin i ego mesto v sisteme sovremennykh mezhdunarodnykh otnoshenii [West Berlin and Its Place in the System of Modern International Relations]* (Moscow: Mysl', 1971), pp. 237–245.
16 *Khrushchev Remembers*, p. 53.
17 *Pravda*, August 12, 1961.
18 *Pravda*, August 8, 1961.
19 *Khrushchev Remembers*, p. 569.
20 This was a phrase used by Khrushchev (ibid., p. 509). The same metaphor occurs frequently in reference to the assumptions of Khrushchev and his supporters in the presidium of the CPSU, in the *Penkovsky Papers*.
21 Brezhnev's report to the 25th CPSU Congress, *Pravda*, February 25, 1976.
22 G. A. Arbatov, "On Soviet-American Relations," *Kommunist*, 3 (1973), p. 105.
23 Ibid.
24 A. O. Chubar'yan, ed., *Evropa XX veka: problemy mira i bezopasnosti [Twentieth-Century Europe: Problems of Peace and Security]* (Moscow: Mezhdunarodnye otnosheniya, 1985), p. 135.
25 Brezhnev at the 23rd CPSU Congress, *Materialy XXIII-ogo s"ezda KPSS [Materials of the XXIIIrd CPSU Congress]* (Moscow: Politizdat, 1966), p. 26.
26 Ibid., p. 27.
27 See, for instance, the analysis by V. G. Vasin, *Godesbergskaya programma SDPG: otkrytoe otrochenie Marksizma [The SPD's Godesberg Program: Open Adolescence of Marxism]* (Moscow: Politizdat, 1963).
28 M. S. Shumskii, *Politika bez budushchego: antikommunizm pravykh liderov SDPG [Policy Without a Future: The Anticommunism of the SPD's Right-Wing Leaders]* (Moscow: Mezhdunarodnye otnosheniya, 1967).
29 M. S. Shumskii, *Vneshnyaya politika i partii FRG [Foreign Policy and the Parties of the FRG]* (Moscow: Mezhdunarodnye otnosheniya, 1961), p. 245. See also Shumskii, *"Vostochnaya" politika FRG: 1949–1966 [The FRG's "Eastern" Policy: 1949–1966]* (Moscow: Nauka, 1967), pp. 292–307.
30 "Soviet Government Declaration on the State of Affairs in the Federal Republic of Germany," *Neues aus der UdSSR*, Soviet embassy, Bonn, February 1, 1967.
31 "Aide memoire of the Soviet Government to the Government of the German Federal Republic on the Question of Renunciation of Force," July 5, 1968, *Neues Deutschland*, July 14, 1968.
32 See, for example, *Neues aus der UdSSR*, Soviet embassy, Bonn, February 1, 1967.
33 *Pravda*, October 19, 1969.

34 *Pravda*, December 4, 1969.
35 See, for instance, V. Yu. Kuz'min, *Vazhnyi faktor stabil'nosti i dobrososedstva v Evrope: sotsialisticheskie strany i FRG v 70-e gody [An Important Factor of Stability and Good Neighborliness in Europe: The Socialist Countries and the FRG in the 1970s]* (Moscow: Mezhdunarodnye otnosheniya, 1980).
36 D. E. Mel'nikov, "The FRG's Foreign Policy," in *Federativnaya Respublika Germanii [The Federal Republic of Germany]*, published in the series *Ekonomika i politka stran sovremmenogo kapitalizma [The Economics and Politics of the Countries of Modern Capitalism]*, p. 463. The paragraph quoted is also the conclusion of the book.
37 B. M. Khalosha, *Voenno-politicheskie soyuzy imperializma: osnovnye osobennosti i tendentsii v 70-kh–nachale 80-kh godov [Imperialism's Military-Political Alliances: Basic Features and Tendencies in the Seventies and Early Eighties]* (Moscow: Nauka, 1982), p. 3.
38 In an interview, jointly with Vadim Zagladin, in *Der Spiegel* (Hamburg), November 5, 1979.
39 F. I. Novik, *SSSR i FRG: problemy sosushchestvovaniya i sotrudnichestva, 1975–1986 [The USSR and FRG: Problems of Coexistence and Cooperation, 1975–1986]* (Moscow: Nauka, 1987), pp. 66–67.
40 Khalosha, *Voenno-politicheskie soyuzy imperializma*, pp. 271–272.
41 Among the plethora of writings see, for instance, V. Anfilov, "The Past Must Not Be Repeated!" *KZ*, June 2, 1984; and Maj. V. Nikanorov, "Indulgence of Militarism," *KZ*, May 18, 1984.
42 See, for example, G. Kirillov and V. Shenaev, "The FRG: The Weakening of the Right-Wing Coalition's Position," in Oleg N. Bykov, ed., *Mezhdunarodnyi ezhegodnik: Vypusk 1985 goda. Politika i ekonomika [International Annual: 1985 Issue. Politics and Economics]* (Moscow: Politizdat, 1985), p. 194. One of the analysts, V. Shenaev, is now deputy head of the newly founded Institute on Europe under the auspices of the USSR Academy of Sciences.
43 For an example that refers to the party leader's warnings, see Col. I. Belov, "Western Europe: the Nuclear Hostage of U.S. Adventurists," *ZVO*, 3 (1984), p. 27.
44 D. E. Mel'nikov, ed., *Zapadnaya Evropa i SShA [Western Europe and the U.S.]* (Moscow: Mysl', 1968), p. 296.
45 Ibid.
46 Kirillov and Shenaev, in Bykov, ed., *Mezhdunarodnyi ezhegodnik*, pp. 193–194.
47 V. Orel, "The Anti-War Movement: Achievements and Prospects," *Kommunist*, 12 (1984), pp. 87–98.
48 *Pravda*, February 26, 1986.
49 See, for example, Yurii Zhukov, "The Anti-War Movements," *IA*, 4 (1987), p. 23.
50 R. A. Faramazyana, ed., *Militarizm: tsifry i fakty [Militarism: Figures and Facts]* (Moscow: Politizdat, 1985), p. 19.
51 Capt. V. Kuzar', "Under the Sign of 'Independent Defense'," *KZ*, July 8, 1984.
52 Col. V. Biryulev, "The FRG's Military-Political Course at the Present Stage," *ZVO*, 8 (1982), pp. 13–14.
53 Col. V. Sergeev, "Mobilization Exercises in the FRG's Air Force," *ZVO*, 9 (1982), pp. 45–46.
54 Capt. S. Chuprov, "In the Spirit of Revanchism and Militarism," *ZVO*, 5 (1982), p. 17.
55 Ibid.
56 Mel'nikov, *Zapadnaya Evropa i SShA*, p. 328.
57 See, for instance, Capt. A. Karemov and Col. G. Semin, "Certain Tenets of the Military Doctrines of the Main NATO European Countries," *ZVO*, 6 (1983), p. 13.

58 L. Borisov, "The Federal Republic of Germany," *ZVO*, 5 (1985), p. 23.
59 Col. G. Semin, "NATO's Military Strategy," *ZVO*, 8 (1983), p. 16.
60 A. A. Kokoshin, "'The Rogers Plan': Alternative Concepts of Defense and Security in Europe," *SShA*, 9 (1985), pp. 5–6.
61 Belov, *ZVO*, 3 (1984), p. 27.
62 Ibid.
63 Maj. Gen. R. Simonyan, "The 'Nuclear Strategy' of the U.S. and NATO," *ZVO*, 6 (1980), p. 9.
64 Belov, *ZVO*, 3 (1984), p. 24.
65 Kokoshin, *SShA*, 9 (1985), p. 10. In 1987, Kokoshin was to give new impetus to Soviet doctrinal thought on strategic conventional defense.
66 V. Lavrenov, "On Certain Aspects of Possible Expansion of the EEC," *MEMO*, 6 (1978), p. 94.
67 Mel'nikov, *Zapadnaya Evropa i SShA*, p. 325.
68 Ibid.
69 On October 20, 1979, on Soviet television, as quoted in Bundespresse- und Informationsamt der Bundesregierung, *Ostinformationen*, No. 203, October 22, 1979.
70 One example of such writing is Daniil Proektor, "Wenn die letzten Limits Fallen," *Neue Zeit* (Moscow), 19 (1984), p. 21.
71 V. Gusenkov, "Contradictions of the Alliance," *Sovetskaya Rossiya*, February 15, 1986.
72 Ibid.
73 Yu. Kovalenko, "A Jubilee to the Sound of Military Marches," *Izvestiya*, January 25, 1988.
74 Ibid.; see also Yu. Kovalenko, "The Brigade, Axis and 'Eurodefense,'" *Izvestiya*, July 22, 1987.
75 Ibid.
76 Lavrenov, *MEMO*, 6 (1978), p. 93.
77 Ibid.
78 S. P. Madzoevskii and S. Skladkevich, "The West European Center: Tendencies in the Development of Mutual Military Ties," *MEMO*, 1 (1978), p. 95.
79 Ibid., p. 96. For a discussion of the role of the WEU, see also G. V. Kolosov, "The Western European Union: Functions and Prospects," *MEMO*, 3 (1975), pp. 139–141. For a discussion of the Eurogroup, exaggerating this organization's significance, see B. M. Khalosha, "The Eurogroup and the Arms Race," *MEMO*, 10 (1976), pp. 98–102.
80 Ibid., p. 97.
81 Kirillov and Shenaev, in Bykov, ed., *Mezhdunarodnyi ezhegodnik*, p. 195.
82 Col. I. Vladimirov, "Basic Tendencies in Military Integration of the West European Countries," *ZVO*, 3 (1982), p. 12.
83 Kirillov and Shenaev, in Bykov, ed., *Mezhdunarodnyi ezhegodnik*, p. 195.
84 See the section of this Chapter "Detente and Its Demise: Emergence of New Threats in the 1970s and 1980s."
85 Vladimirov, *ZVO*, 3 (1982), p. 13.
86 Kuzar', *KZ*, July 8, 1984.
87 Madzoevskii and Skladkevich, *MEMO*, 1 (1978), p. 95.
88 Ibid., p. 99.
89 Ibid., p. 98.
90 Vladimirov, *ZVO*, 3 (1982), p. 15.
91 Madzoevskii and Skaldkevich, *MEMO*, 1 (1978), p. 99.
92 Kirillov and Shenaev, in Bykov, ed., *Mezhdunarodnyi ezhegodnik*, p. 195.
93 Vladimirov, *ZVO*, 3 (1982), p. 16.

94 Kirillov and Shenaev, in Bykov, ed., *Mezhdunarodnyi ezhegodnik*, p. 195.
95 V. Gusenkov, *Sovetskaya Rossiya*, February 15, 1986.
96 The nuclear aspects of Franco-German military cooperation are discussed under the heading "Quest for the Acquisition of Nuclear Weapons?"
97 Kovalenko, *Izvestiya*, July 22, 1987.
98 Viktor Levin, Radio Moscow, in Russian, January 22, 1988.
99 *Bundeswehr aktuell*, October 12, 1987. Falin is now head of the *Novosti* news agency and a member of the CPSU Central Committee.
100 *Der Stern*, April 2, 1987.
101 At a CDU conference on foreign policy on April 14, 1988, as quoted in *Frankfurter Allgemeine Zeitung*, April 15, 1988.
102 "East-West: How to Approach Each Other," *Moskovskie novosti*, 9 (1988), p. 6. Portugalov replied to an article by West German Major General Hüttl in the same issue of the journal.
103 A. Yu. Mel'vil', "The 'Image of the Enemy' and New Political Thinking," *SShA*, 1 (1988), p. 34.
104 Ibid.
105 Ibid., pp. 30, 39.
106 *Pravda*, January 16, 1988.
107 Ibid.
108 V. Zhurkin, S. Karaganov, and A. Kortunov, "Challenges to Security—Old and New," *Kommunist*, 1 (1988), pp. 43–44; similarly innovative is their article, "On Reasonable Sufficiency," *SShA*, 12 (1987), pp.11–21.
109 Zhurkin et al., *Kommunist*, 1 (1988), pp. 43–44.
110 Ibid., p. 44.
111 Ibid.

9
Soviet Public Diplomacy toward West Germany under Gorbachev[1]

Robbin F. Laird

This chapter describes Soviet analyses of developments in the Federal Republic of Germany (FRG) during the 1980s. These analyses are then related to the Soviet public diplomacy effort in West Germany. Finally, West German reactions to this Soviet public diplomacy effort are evaluated.

Specific Soviet Perceptions of West German Developments

The evolution of Soviet public diplomacy toward West Germany illustrates the general trends in Soviet public diplomacy that are identified in chapter 1 of this book. However, the Soviets have formulated their West German policy in accordance with specific perceptions of developments in the FRG; the following pages identify some specific perceptions evident in the Soviet literature published in the 1980s.

The most salient background information for understanding the Mikhail Gorbachev administration's opening to West Germany was the previous administration's handling of the intermediate-range nuclear forces (INF) issue and the German role in that issue. Andrei Gromyko and Yurii Andropov tried to isolate the FRG and to play on its peace movement and left-wing parties in order to block the North Atlantic Treaty Organization's (NATO)

INF deployments. This policy failed, and the Soviets linked their failure with the emergence of "revanchism" within West Germany. Revanchism meant the return of militarist tendencies in the FRG and with it a resurgence of Atlanticism in West German policy.

For the Gorbachev administration, the further development of both revanchism and Atlanticism in West German policy had to be deflected. In addition, in the wake of the Reykjavik summit another troubling trend emerged: the growing commitment to Europeanism in West German policy. The conservative governments in Great Britain and West Germany, along with the cohabitation government in France, resisted denuclearization efforts by the superpowers and became bulwarks against change in NATO's security policy.

West German Foreign Policy

According to Soviet analysts, the Europeanization of security policy has been used increasingly by the West Germans to justify an enhancement of their role in the Western Alliance. In this connection, the West Germans have emphasized the importance of activating the Eurogroup, the Independent European Program Group (IEPG), and the Western European Union (WEU). "The [FRG] ruling circles intend to use the slogan of 'Europeanization' of arms in order to strengthen the FRG's position in NATO."[2] West German leaders hope to make Europe more competitive with the United States, but in order to strengthen the European pillar rather than to eliminate the United States from European defense altogether.[3] Finally, West German leaders also use Europeanization as a means of enhancing West German military capability: "In addition to financial benefits (due to the pooling of resources) Bonn's rulers hope that intensified cooperation within Western Europe would be essentially useful in providing a cover for their own commitment to continued build-up of the FRG's military potential by dissolving it in the 'all-European responsibility' and 'all-European tasks.'"[4]

Soviet analysts perceive West German policy to be central to the further development of Alliance policy, thereby justifying the development of a new opening to the FRG under Gorbachev. "Will the long-awaited turn towards detente occur in Europe, or will tensions continue to be built up here? The answer . . . largely depends on the stance assumed by the Federal Republic of Germany on the major international issues." The analyst added that "the myth . . . that the West German economic giant is a political dwarf in its influence on European and world processes, has definitely become a thing of the past. Today it is impossible to imagine a West European position being elaborated without taking the stance of the FRG into account."[5]

A number of objective factors also provide the basis for West Germany's supporting a detente policy. Its geostrategic location, the large military presence in the country, and Bonn's lack of control over nuclear and chemical weapons on its territory "prompt Bonn to be cautious in politics and to pursue a course toward preventing a war and maintaining normal conditions for trade and economic cooperation, which, in fact, amounts to a policy of peaceful coexistence."[6] These objective interests provide for the possible confluence of Soviet and West German policies. As L. Borisov comments, "In recent times the Bonn leaders have been making statements which are largely consonant with Soviet foreign-policy philosophy."[7] Especially important is the West German belief that security cannot be obtained through military power alone, but that there is an objective need for peaceful coexistence. While the position of Foreign Minister Hans-Dietrich Genscher is well documented (that West Germany should use "the new opportunities offered today by Soviet foreign policy"), the Soviets also hasten to point to Chancellor Helmut Kohl's speech on March 18, 1987, in which he stated that the West German government should "pursue 'an active and all-embracing policy of peace and good neighborly relations.'"[8]

Soviet–West German Relations

There is a rich Soviet literature that assesses the Soviet relationship with the Federal Republic of Germany. The thrust of this literature emphasizes that West German policy toward the Soviet Union has been divided into four phases since 1970. The first phase, associated with the Willy Brandt government, was one of building detente with the Soviet Union. This positive experience has led the Soviets to continue to hold Brandt it high esteem and to treat his visits to Moscow as important opportunities for highlighting the possibility of positive developments in West German–Soviet relations.

The second phase, associated with the Helmut Schmidt government, coincided with a more contradictory detente policy. While the coalition of the Social Democratic Party (SPD) and the Free Democratic Party (FDP) pursued detente with the Soviet Union, the coalition simultaneously worked with the Americans to strengthen Alliance military policy. As one Soviet analyst put it, "The FRG's entire policy in the 1970s developed along two paths. On the one hand, the FRG expressed an interest in strengthening detente, and on the other, it did not for a minute cease to build up its military potential and to equip the *Bundeswehr* with modern weapons."[9] At least part of the reason for this contradictory policy was due to pressure from the political right, including the Christian Democratic Union (CDU)–Christian Social Union (CSU) leadership, who demanded support for U.S. policies.[10]

The third phase of West German policy coincided with the election of the new coalition government led by the CDU. This phase was dominated by confrontation with Moscow and the battle over INF. Soviet analysts noted an important difference between CDU promises and practices: "When the conservative Christian Democratic Union/Christian Social Union parties came to power in Bonn in 1982, they were not lacking in promises to preserve continuity in the country's foreign policy. Now it is 1984. The two intervening years have brought increasing evidence to the contrary."[11]

The basic judgment on this third phase was that the problem lay in West Germany's acquiescence in U.S. policy. Notably, the Kohl government's decision to deploy INF and support the militarization of space

> represents a direct challenge to the security of the Soviet Union and other socialist countries. Attempts by certain figures in the ruling coalition, including members of the cabinet, to put into doubt the political and territorial realities in Europe, to distort the essence of the socialist states' foreign policy also disrupt the development of normal relations. At the same time, the FRG government cannot fail to take into account the opinion of a majority of the population, which advocates mutually beneficial cooperation with the socialist countries.[12]

Thus, Bonn was seen to adopt the American approach in many aspects of its relations with the USSR, especially in security matters.[13]

But even during the third phase, West German conservatives were unable to jettison the detente policy. According to G. Kirillov, "One is bound to conclude that the relations between the USSR and the FRG have become noticeably more complex in the 1980s as compared to the 1970s. But this is not to say that they have come to a standstill. The impetus given to the Soviet–West German relationship by the Moscow Treaty is still at work. The two states continue political contacts at various levels. The two countries maintain a relatively high level of trade and economic ties."[14]

The fourth phase has been characterized by a reopening to the Kohl government. The regeneration of the detente process with the West Germans, although difficult, is necessary for the Soviets in order to cope with developments in the Alliance as a whole. For the West Germans, their interest in East-West trade has been especially important in maintaining support for detente. Thus, even during the deterioration in East-West relations, the West German government "sought to continue cooperation with the socialist countries in the economic area.... The tendencies toward reviving the trade, economic and other ties between the FRG and the socialist countries testifies to the fact that the FRG did not want to throw away completely the positive experience of cooperation in the 1970s."[15] The author adds that the broad public support for *Ostpolitik* has become a barrier to conservative efforts to undercut relations with the Soviet Union: "On the whole, the development of events in the first half of the 1980s has shown that the *Ostpolitik* begun by the Moscow Treaty has put down deep roots and contrary to the actions of the rightists and the pressure from the outside, it is finding support from the overwhelming majority of the population of the Federal Republic of Germany."[16]

After a period of confrontation, the Soviets began to repair relations with Kohl's government. During this period, the focus has been on common interests and "practical" problems that the two governments can solve together. For example, the Soviet ambassador to Bonn, Yulii Kvitsinskii, made an interesting comparison between the state of current Soviet-FRG relations and those that existed during the period from 1978 to 1981, when he was last in Bonn. The 1978–1981 years "were productive ones for our relations.... We are now undergoing a more complicated period in international life and, of course, it reflects on the relations between our states. I suggest concentrating on improving the current state of affairs. Our relations are worthy and capable of improvement. I believe that much, and much good, can be done and that in no case must [the] substance of our relations deteriorate."[17]

Foreign Minister Genscher's visit to Moscow in July 1986 was used by the Soviets to highlight the new possibilities for cooperation with the FRG. Genscher delivered a message from Kohl to Gorbachev in which the chancellor stressed "the readiness of the Federal Government to strengthen dialogue with the Soviet leadership and to improve relations in all fields."[18] The Soviet press responded favorably to Genscher's comment that there were "many unused opportunities" for more cooperation, including economic ties.[19]

Nonetheless, the reopening of the Soviet-West German dialogue was stalled by Chancellor Kohl's remarks comparing Gorbachev's new thinking to the propaganda of Nazi leader Josef Goebbels. As Foreign Minister Eduard Shevardnadze commented:

> Fair relations have formed between us and the FRG in the sphere of economic cooperation and political contacts, including at a fairly high level. But recently certain remarks of an insulting nature were made by the Federal Chancellor with regard to the Soviet leadership which have angered us to the depths of our souls. These remarks have been condemned also by public opinion in the FRG itself. As far as the people of the FRG, business circles and corresponding parties and organizations are concerned, we are in favor of normal mutually advantageous relations.[20]

Similarly, a front page editorial in *Pravda* had the following reaction to Kohl's Goebbels statement:

> These pronouncements made by Kohl, which are unprecedented in the history of Soviet-West German relations, and even in all relations between states maintaining diplomatic relations, have given rise to profound indignation everywhere, including the FRG. They are staggering in their irresponsibility.... A deep shadow has been cast on Soviet-West German

relations. They have been poisoned, and this at a time when they had begun to take on a new dynamism and when their further development could have improved in a most salutary way the situation in Europe and in the world as a whole. Understandably, in this kind of atmosphere a series of planned meetings and visits on a bilateral basis could not take place. The Soviet side has earlier demonstrated the maximum of restraint and tolerance, but there is a limit to everything.[21]

The reelection of the conservative-led coalition government in early 1987 stimulated Soviet attempts to reenergize Soviet-FRG relations. Kohl was quoted as saying that relations with the Soviet Union are of "central importance" to the FRG.[22] The new government was seen to be interested in enhancing political dialogue, cooperating in scientific-technological and environmental areas, and improving humanitarian, cultural, and economic relations. Also underscored was Genscher's statement that since Reykjavik new prospects have opened up for East-West relations, including in the area of disarmament.[23]

Yet Soviet analyses still tended to emphasize continuing problems in dealing with the Kohl administration. A June 1987 assessment of bilateral relations is typical in this respect:

For a number of years, which have made history as the period of detente relations between statesmen, politicians and mass organizations of the USSR and the FRG and trade and economic contacts between the two countries developed successfully and cultural and scientific cooperation between them also broadened. Lately, however, the state of things in the field has changed, not at all for the better. "Discussions" have resumed in the FRG on revising territorial and political realities in Europe and revanchist organizations stepped up their activities with the knowledge and connivance of authorities. Revanchists' gatherings are often attended by high-ranking representatives of the ruling coalition, including the Chancellor himself. Bonn's approach to such an important issue of the times as disarmament cannot be called constructive either.[24]

When President Richard von Weizsaecker went to the Soviet Union in mid-1987, Gorbachev noted the necessity for progress in Soviet-FRG relations but that the slide into confrontation had at least stopped. Speaking with reference to Genscher's visit to Moscow the previous year, Gorbachev noted that "agreement seemed to have been reached on 'opening a new page' in relations between the two countries. However, to date, it has remained unfilled, and at one time there was even a threat that it would be shut. Fortunately, this did not happen."[25]

General Soviet commentaries on Soviet-FRG relations after the von Weizsaecker visit noted an improvement in the Bonn government's position. Although changes in rhetoric are not conclusive, "Still, these statements must not be ignored. After all, in the past all the CDU/CSU governments invariably followed the rule that it is better to back up a tough line in relations with the East and build up arms than to pursue a policy of detente and disarmament."[26] This Soviet analysis then underscores Gorbachev's comment that stable Soviet-FRG relations ensure European stability; poor political bilateral relations over the past few years are blamed on INF deployment and the accompanying revival of militaristic forces in West Germany.[27]

When former CSU leader Franz Josef Strauss visited the Soviet Union in December 1987, the Soviets conveyed their willingness to reach out to all political factions in the FRG in their efforts to improve relations. For years the Soviet press had identified Strauss as the archenemy of detente and a political force to be avoided. From this perspective, Gorbachev's comments during a two-plus-hour meeting with Strauss are notable. According to the Soviet press, Gorbachev "expressed the hope that the new year 1988 would be marked by more intensive and constructive political dialogue between the USSR and the FRG, and broader contacts and meetings between the two countries. Both sides showed understanding of the importance of making Soviet–West German relations a matter of practical policy."[28]

Since the Strauss visit, the Soviet press has noted the growing possibilities for cooperation with the FRG. Although the main emphasis in the Soviet dialogue with the West has shifted to improving relations with the United States, the visit by Kohl to Moscow in 1988 and the return visit by Gorbachev to Bonn in 1989 were designed to connect the reopening to Bonn with the broader detente policy toward the West. The previous barriers are perceived to have been removed, namely, the INF deployment, the "revanchist" threat, and the unfortunate remarks by Kohl (although the latter are not completely forgotten).

West German Economic Policy

Soviet commentary on West German economic policy includes three principal themes. First, the West German economy is the most important one in Western Europe. Second, the weight given economic considerations is consistently the most important factor in shaping overall West German government policy. And third, the FRG is the most important Western trading partner of the Soviet Union.

Soviet assessments of the West German economy underscore its significance for the FRG's overall power position in Western Europe. "West German militarism operates on a powerful economic and technological base that does not have an equal within Western Europe. The FRG's GNP [gross national product] is three-fourths the combined GNP of Great Britain and France. In terms of industrial production, it exceeds Great Britain and France taken together. In currency reserves, its superiority is even greater. In terms of the FRG's competitiveness,... it is fourth place... in the capitalist world, far surpassing France (19th place) and Great Britain (14th place)."[29] The Soviets consider West Germany to be "the economic nucleus of the West European imperialist 'power center.'" And it uses its economic lever to influence the policies of other Western states.[30]

The desire of West German leaders to maintain economic growth and development provides them with an incentive for cooperating in arms control efforts. According to Soviet thinking, the economic burden of the defense budget makes it impossible for the West German state to finance the arms race through normal economic growth rates.[31] An additional incentive for reducing the military confrontation with the Soviet Union is based on the West German role in East-West trade. Typically, the Soviets note that the FRG "has long been the largest trade and economic partner of socialist countries among the capitalist states."[32]

The Soviets further underscore the important role that West German businesspersons play in encouraging their government to increase East-West trade. For example, Nikolai Ryzhkov noted: "We know that there are business circles in the FRG which are interested in closer economic ties with the USSR. Such ties have indeed become traditional." Ryzhkov continues, noting the conflict between West German interest in East-West trade and the subordination of that interest to American definitions of Alliance security interests:

> But at the same time, the FRG is following in the wake of the American policy of bans and embargoes. In our opinion, the purpose of such proscriptions is not to prevent the USSR and other socialist countries from receiving up-to-date technology of military or dual purpose. It would be naive to assume that we would wish to import some elements for our military technology from West European countries. As we see it, by imposing such bans the U.S. is raising obstacles to the expansion of economic ties between the socialist and capitalist countries of Europe. Whether this is of advantage or disadvantage to Western Europe is for you to judge.[33]

In short, the economic interests of the FRG are a critical dimension in the development of Soviet-West German relations. Such trade can provide the basis for developing joint interests and can be useful in building shared perspectives on other problems as well as in East-West relations. In this period of *perestroika*, the Soviets have clearly articulated the desire to see the West Germans participate more fully in the economic revitalization of the Soviet Union. West German participation in this process is seen to encourage other Western countries to follow suit as well.

The West German Bundeswehr *and the Military-Industrial Complex*

If economic interests drive Soviet and West German policy into a cooperative effort, West German "militarism" remains a key force impeding detente. The West German armed forces are the foundation of NATO's conventional force operations. The Kohl government has also emphasized solidarity with the Americans in enhancing nuclear force modernization (in the case of INF) and a new aggressive effort in strategic weapons (through participation in the Strategic Defense Initiative [SDI]). Also, the West Germans have sought to enhance their de facto involvement in Alliance nuclear issues, especially with regard to the evolving Franco-German relationship.

Soviet analysts consider the *Bundeswehr*, along with American troops in the FRG, "to make up the main strike force" of NATO in Europe. The FRG army is also considered to be the basic strike force of NATO ground troops in the central European theater of military operations (TVD).[34] The Soviet emphasis on the primacy of the West German role in conventional forces makes the West Germans the prime target for efforts in conventional arms control policy. If the Soviets can reduce the salience of the conventional military confrontation, then broader prospects for cooperation latent in the West German–Soviet economic relationship might be realized. At the same time, Bonn relies on the primacy of these forces in maintaining its influence in the Western Alliance. "The Bonn government apparently sees the constant development of its military muscle and the expansion of its military-economic activity as one of the most important ways of strengthening the FRG's position in NATO."[35]

Because the FRG cannot have direct access to nuclear weapons, it uses its conventional military power to gain influence in NATO. One Soviet analyst cites a statement by Kohl that increased *Bundeswehr* capabilities will enhance the military and political clout of West Germany in the Alliance. Kohl further noted that without a powerful *Bundeswehr*, West Germany would not have been able to influence NATO decision-making effectively. Thus "the Bonn government openly links the role and international weight of West Germany with the strength of its armed forces, with the growth of the country's military potential."[36]

Nonetheless, West Germany has linked its military strategy closely with that of the Americans; this means that the fate of the *Bundeswehr* is tied up with the Americans. "The FRG's military preparations significantly exceed its defensive requirements, but they are organically incorporated into American militaristic plans, and this means that West Germany, by the will of its military-economy and political elite, is made degradingly dependent on the aggressive strategy of the American ruling elite."[37] It should be noted, however, that influence can flow both ways. If the Soviets believe the fate of NATO's conventional operations rests heavily on the U.S.-FRG military relationship and that the *Bundeswehr* is critical to that relationship, breakthroughs in relations with the FRG in security policy could be significant in influencing Washington.

Aside from West German conventional forces, there is the issue of nuclear weapons. Despite some statements that the FRG does not want to acquire nuclear weapons, it really is looking for ways to get its hand on the button, according to several Soviet military analysts.[38] In this vein, Captain V. Kuzar' notes that the Tornado is to be used mainly for "delivering nuclear strikes against precise targets at low altitudes."[39]

Soviet analysts consider the impact of Bonn's military preparations on the East-West situation to be profound. For example, one analyst argues that

> the present course of the Bonn ruling circles towards a massive buildup of the Bundeswehr arsenals meets the interests of bellicose forces in the USA and NATO, and is in conformity with the revanchist aspirations of those on the Rhine who cannot forget the past. At the same time, it poisons the political climate, undermines trust and mutual understanding among European nations, and runs counter to the interests of peace and the security of the continent's countries, including the FRG itself.[40]

In addition to the deployment of forces, the FRG has become a prime producer of military weapons. In fact, the West German military-industrial complex is a prime mover in the effort to Europeanize the productive base of the Alliance. "The FRG is involved in a far larger number of joint military programs than other West European countries. Its involvement in joint programs enables Bonn to bypass established limitations and gain access to the latest technology and know-how in areas where West German monopolies... lag behind their NATO partners."[41]

In short, the military foundation of West German security policy limits the Soviet Union's ability to establish a genuine breakthrough in Soviet-West German relations. It is necessary for the Soviets to reduce West German threat perceptions and perhaps to adopt a genuinely defensive policy in order to develop real possibilities for breakthroughs in East-West security policy.

West Germany and INF

Throughout the deployment crisis, the Soviet assessment of the INF issue relied on the significance of the political dimensions of the controversy. West Germany was seen to be a major player in the deployment process, and such participation led, in the Soviet view, to the encouragement of negative tendencies in West German policy. For example, one Soviet analyst commented that "having agreed to deploy American Pershing and cruise missiles in the country, the West German government acted, in effect, as one of the main initiators of the realization of NATO's decision to 'rearm'."[42]

The Alliance action led to the reversal of detente, which in turn led to a resurgence of revanchism in West German policy. "Detente was hit hard when new U.S. nuclear first-strike weapons began to be deployed in West Germany with the consent of the FRG government, in keeping with the will of the U.S. ruling elite.... Revanchism and the missile deployment appeared to be communicating vessels."[43] Lest one wonder what constitutes revanchism in this context, the author later explains that "as a result of the deployment of new U.S. nuclear missiles on West German soil, forces are again rearing their heads in the FRG which are intent on calling in question postwar political and territorial realities in Europe."[44]

The antirevanchist campaign encouraged the Soviets to follow a path of intimidation rather than cooperation. This is why it has been so important to the Soviet leadership to put the INF issue behind them in order to relaunch detente in Western Europe. "A necessary condition for stability in the positive processes in Europe and in other regions is a respect for the territorial-political realities which came about as a result of the Second World War. The CPSU ... will quash any appearance of revanchism." Tsvetkov concluded that "West German militarists have chosen as their main political course revanchism and aggression, giving birth to adventuristic plans for the violent changing of national borders in Europe."[45]

Nor has the West German action in supporting the deployment of American missiles enhanced the security of the German people, according to the Soviet argument. The INF deployment means that "West Germany is thus becoming a launching pad for American first-strike nuclear missiles, which means that for the first time since the war a threat to the Soviet people and Soviet allies will be emanating from German soil."[46] West Germany's loss of control over the fate of its own security (because of its participation in this deployment) was also an important element of Soviet analysis. By accepting INF missiles, Bonn is "creating a situation when a new war could be launched from the territory of the FRG even without the knowledge of its authorities."[47]

Despite supporting NATO's zero option, Kohl seemed to change his position after the Reykjavik summit. He began to call for progressively lowering the number of weapons but did not embrace their complete abolition. The West German government was, in fact, divided over what to do. For Genscher, elimination of these missiles represented an "unprecedented epoch-making step whose implementation would ensure security for Europe and would positively influence East-West relations." In contrast, Defense

Minister Manfred Woerner, while not rejecting the zero option completely, did caution about Soviet superiority in conventional weapons.[48]

The Soviets decided to try to meddle in West German politics by insisting that the Pershing IAs be included in the treaty, given that the warheads are under U.S. control. The Soviet argument was that "nuclear warheads for these systems are deployed on West German territory but are under the control of the U.S. military command. It goes without saying that the question whether it is nuclear or conventional warheads that will be used on these missiles will be decided primarily by the USA."[49] To place pressure on the West German governing coalition, the Soviet leadership raised the question of whether West Germany intended to acquire an independent nuclear capability and thereby vitiate the nonproliferation treaty. For example, Shevardnadze stated: "If, as certain American leaders and their West German partners pretend, the Pershing IAs belong to a third country, let us ask once again why and with what right does West Germany have nuclear weapons." If West Germany was hoping to acquire nuclear weaponry it would be creating a "world political crisis." Such an action would jeopardize the Soviet-U.S. talks and create a situation in which "our allies could demand the installation of similar systems on their territory and the Soviet Union could respond to their demand." Furthermore, "Either the Pershing IAs belong to the United States, and should be included in the INF negotiations, or they belong to West Germany, which has neither the legal nor moral right to possess them."[50]

Soviet analysts argued that divided opinion within the West German government reinforced by the absence of strong support from its allies led the FRG to accept inclusion of the Pershing IA warheads in the agreement.

> Only after a long period of procrastinations and heated debate within the ruling coalition, was the FRG government, fearing to be in isolation even among its NATO allies, compelled to back up the USSR-proposed "zero option" on missiles with a range between 500 and 5,500 kilometers. However, ... it demanded that its 72 Pershing IA missiles with a range of 750 km and the U.S. nuclear warheads for them be preserved. These systems, it claimed, like the nuclear potentials of Britain and France, should not be counted at the Geneva talks. But, first, if the warheads to West German Pershings ... are not under U.S. lock and key, as they were believed to be before, but are under FRG control, then it is a gross violation by both countries of the Treaty on the Non-Proliferation of Nuclear Weapons. If they are under U.S. control, then the warheads should be destroyed together with all the other Soviet and American warheads for medium-range and enhanced-range tactical missiles in Europe. What would be the reaction in the FRG if the USSR's allies had similar carrier missiles and the nuclear warheads for them were stored close to the launching positions in Soviet depots? In this case, perhaps, Bonn

would look at the issue of third countries' weapons in a different way.⁵¹

West German-American Relations

The resolution of the INF issue has been perceived by the Soviets to be central to further progress in Soviet-West German relations. This is due above all to the role nuclear weapons have played in binding West Germany to American interests. By breaking this connection, West Germany's room for maneuver will be enhanced. Illustrative of this logic is the following: "The Kohl-Genscher government is turning the country from an equal partner into the U.S.' faithfully servile assistant. National interests are being sacrificed for solidarity with the Reagan administration. This was manifested above all in the deployment of American medium-range missiles in the FRG's territory."⁵²

Nonetheless, the Soviets understood that even in a nuclear buildup, West German and American interests were different and could be exploited. "This country's government is not, of course, ecstatic about U.S. plans to turn its territory into a theater of military action, including 'limited' nuclear actions, since it understands that a U.S. attack on the USSR and its allies will lead to catastrophic consequences for West Germany." In addition, FRG ruling circles are seeking "greater equality" in American-West German relations.⁵³

By the time of the January 1987 election, the issue of conflict between the United States and the FRG had become a central one, according to Soviet analysts. This election was interpreted basically as a clash between the supporters and opponents of pro-U.S. policies. Kohl's pro-U.S. position was one of the reasons for the loss of CDU/CSU votes in the election. Soviet analysts cite public opinion data indicating that West Germans overwhelmingly favor a detente policy toward the Soviet Union.⁵⁴

Two themes regarding the party positions on U.S.-FRG relations and their effect on East-West relations have been especially noted over the past 18 months. First, frequent disagreements within the governing coalition have been noted. There is manifest discord between the CSU and some right-wing CDU members on the one hand, and the FDP on the other hand, about *Ostpolitik*, disarmament, and arms control.⁵⁵ Second, the SPD's utility has been seen to increase now that it has reentered the political mainstream, given the promise of change in East-West relations associated with the INF treaty. "The Social Democrats recognize the 'central importance' of relations with the U.S. for the FRG. They do not deny the need to strengthen the alliance with the U.S. and the FRG's position in NATO; however, they do place a greater stress on national and West German interests than the CDU/CSU does." They do not question the FRG's adherence to NATO, but they do want the FRG and all of Western Europe to be more independent from the United States and to be respected partners with the East.⁵⁶

The Soviet Approach to West German Elites

Gorbachev's public diplomacy toward the FRG has been designed to address the disquieting trends in West German security policy. Revanchism is to be deflected by encouraging the West Germans to return to their detente policy of the 1970s, a policy that accepted the territorial and political status quo in Europe. To accomplish this, Gorbachev has emphasized a new detente with the Alliance in which the Soviets accept the "realities" of the foundations of Western security policy in exchange for a recognition of the "realities" of Eastern security policy.

To reduce militarism in West German policy, the Soviet leadership has stressed the possibilities of serious reductions in the Soviet military threat. In fact, the goal is to eliminate as much as possible the ability of the West German conservatives to use the excuse of the Soviet military threat to justify West German military expenditures. Furthermore, the Soviets have emphasized the possibilities of shifting East-West military policy to a defensive basis, a policy that would allow the SPD to return to the center of the political spectrum on security issues.

To reduce Atlanticism and Europeanism, the Soviets emphasize the primacy of the common European home, but one without European military integration. The emphasis on this theme suggests a distancing from the Americans, although the new realism in Soviet policy recognizes the durability of the NATO Alliance and the legitimacy of some form of American military presence in Europe. The goal is simply to reduce the cohesion of the Alliance and the nature of the U.S. military presence.

But the Soviets do not want to see a reduced American presence lead to an enhanced European role in the Alliance, a process that could occur only with vigorous West German involvement. The Soviets emphasize the legitimacy of the development of West European economic and even political integration (some Soviets even say they accept a West European role in Eastern Europe) but deny the legitimacy of West European military integration. According to the Soviets, the common European home is possible only if there is no military integration of Western Europe.

Prior to the Gorbachev administration, the Soviets targeted their efforts on groups opposed to the mainstream policies of the FRG, especially the peace movement and the SPD (after the latter's change in policy during the INF crisis). Under Gorbachev the Soviets have shifted their approach, now dealing directly with centrist groups, including prominent politicians whom the Soviets have hitherto considered anathema.

The Soviets foster a dialogue with these centrist groups by openly discussing issues and publicly demonstrating a self-critical attitude. When U.S. and Soviet representatives appear at common forums in the FRG, the Soviets focus on the political context of security policy, whereas the Americans are more likely to discuss military-technical issues. The Soviets appear to be more interested in peace than the Americans, precisely because military-technical discussions are so rare in the West German public domain that the mere discussion of security in these terms shows a blatant American misunderstanding of how to influence the West German public. The Soviet approach, on the other hand, reveals a greater sensitivity to West German opinion.

Historically, the Americans' ability to influence West German opinion has been aided significantly by their candor in discussing issues and admitting previous mistakes. The Soviets are seeking to eliminate this advantage. In addition, by developing a dialogue with mainstream groups, Soviet discussions with other groups—such as SPD politicians critical of centrist politics—have become less marginal. Discussions with the SPD have become just one element in West German–Soviet discussions rather than attracting all the attention. Finally, the Soviets have related discussions of reasonable sufficiency to West German concepts of alternative defense. To the extent that the reasonable-sufficiency concept becomes the coin of the realm in security issues, alternative defense concepts become less marginal and gain respectability and plausibility.

The Soviet attempt to influence West German elite opinion is well adapted to that country's realities. Whereas in the United States, Soviet spokespersons frequently appear as commentators on news programs, the operation of the West German media makes such an approach difficult. It is much more important to work directly with the network of elites in discussion groups and forums, as well as to try to affect the positions of West German journalists. The West German print media are very influential in shaping elite understandings, much more so than is television.

Typical of the Soviet effort to participate in meetings and roundtables was the invitation extended to the *Bergedorfer Gesprachskreis* group to visit Moscow in March 1987. Among the participants in this group were Egon Bahr (SPD), Horst Teltschik (advisor to Chancellor Kohl), Lothar Ruhl (state secretary in the defense ministry) and former Chancellor Helmut Schmidt. It is a politically bipartisan association comprised of important figures from West German politics, economy, and society, which meets on an ad hoc basis to discuss critical contemporary issues. The Soviets invited this group to Moscow and discussed their views on *perestroika* and new thinking as well as on the actions the Soviets have already taken to implement these changes.

The Soviets actively participate in conventions, meetings, and forums held in the FRG as well. For example, Ambassador Kvitsinskii participated in a 1987 conference on foreign policy that was organized by the CDU. At this meeting Kvitsinskii emphasized that the Soviet Union was willing to alter its East-West policy, especially its position on the presence of the United States in

Europe. He even suggested that the American military presence in Western Europe was legitimate; the Soviet dissatisfaction with this presence centered merely on how the Americans used their presence to influence progress.

Also typical of Soviet activities was a discussion by Valentin Falin, former Soviet ambassador to the FRG and current head of the Soviet press agency Novosti. Early in 1988 Falin held a press conference in West Germany in which he urged the West Germans not to rush ahead with short-range nuclear modernization. He asked the West Germans to give the Soviet Union time to realize the promise of *perestroika* and not to force the Soviet leadership to react to a new round of arms modernization on West German soil. He suggested that the government build on its responsible performance in resolving the Pershing IA problem. But if the West Germans ignore this advice, of course, the Soviet Union could harden its policy toward the FRG. Thus, Falin combined an effort to woo West German elites with a clear warning. This combination is typical of most Soviet efforts in dealing with West German elites, although the attempt to woo them currently predominates.

Kvitsinskii and Falin are the most visible representatives of the Soviet Union in the FRG. Both speak fluent German and are adept at presenting the proper image. Of the two, Kvitsinskii has become the more visible figure. He is perceived to be an ambitious man who is on the rise and is therefore influential with Gorbachev. Contrary to his predecessor (Vladimir Semyonov), he is perceived to have much greater authority in Moscow. While Semyonov was seen to represent the "old way of thinking," Kvitsinskii is strongly in favor of the new political thinking, of *glasnost'* and *perestroika*. While in American eyes he has been first and foremost an arms control specialist, he belongs as well to the so-called Germanisty. He is a protégé of Politburo member Aleksandr Yakovlev and a close friend of Nikolai Portugalov (a senior Soviet Germanist) and Valentin Falin. He is definitely not an *Amerikanist*, which means that in the International Department of the CPSU he is linked more to Vadim Zagladin than to Anatolii Dobrynin. He also has close connections to TASS and is a friend of the chairman of the Gesellschaft BRD-UdSSR, Sergei Losev, who is also affiliated with TASS.

Kvitsinskii is not only a candidate member of the Central Committee but is also one of the relatively few foreign policy specialists given the opportunity to speak before the Central Committee. Prior to Kvitsinskii's tenure, it was impossible to contact people from the Central Committee apparatus via the Soviet embassy. Now the embassy acts as an intermediary agency with the Central Committee apparatus as well. Soviet embassy officials under Kvitsinskii are given relatively wide latitude in dealing with West German elites and have tried to use their embassy as the Americans use theirs, as a place in which frank discussions can be held. For example, a leading West German specialist on the Soviet Union who is a CDU official was invited recently to the embassy to give a lecture on the origins of World War II.

Falin remains an important player as well and has used his position as the head of Novosti to promote Soviet public diplomacy in West Germany. Most notably, he was the guiding force behind the decision to publish a West German version of *Moscow News*. This paper was first published in 1988 and is not a German translation of the Russian-language paper. Rather, it is a monthly publication of the most prominent examples of *glasnost'* and new thinking. The paper does not focus on German-Soviet relations but rather deals mainly with Soviet domestic policy issues associated with *perestroika*.

The German version of *Moscow News* is written in a chatty style. The Soviets want to use this publication to suggest subliminally that press freedom has been significantly expanded in the Soviet Union (most West German readers of *Moscow News* do not realize that it is not a literal translation, nor do they realize how limited the circulation of the Russian-language version of *Moscow News* is in the Soviet Union). By means of an English title and use of the word *news*, the Soviets are trying to convey to the reader that this publication is something akin to an American newspaper.

There is a curious mixture of old thinking among the German hands in the Ministry of Foreign Affairs and new thinking in the International Department of the Central Committee. West German officials at the working level continue to experience frustration in dealing with their Soviet counterparts. For example, the former challenge the recognition of West Berlin in negotiating the bilateral agreement between the European Economic Community (EEC) and the Council for Mutual Economic Assistance (CMEA). In contrast, International Department officials behave somewhat contradictorily, thereby limiting the prospects of successfully influencing West German elite opinion. Most notably, West Germans see a consistent anti-German feeling among high-ranking Soviet officials (excluding Gorbachev). For example, during Dobrynin's first visit to West Germany in 1987, SPD officials clearly discerned anti-German feelings on his part, although by the end of his stay he was seen to demonstrate greater openness to West German concerns. Nevertheless, the speech Dobrynin delivered to the SPD also illustrated the limits to Soviet innovation, for it was a speech about the U.S.-Soviet relationship that would have been more appropriate in Washington than in Bonn.

Since the Strauss visit to Moscow in December 1987, Gorbachev has been conveying his desire to accelerate progress in the bilateral relationship. During the visit he was frank and open with this representative of "revanchism." Since then, he has been meeting with virtually every prominent West German visitor to Moscow. He met for 4 hours with Willy Brandt in April 1988 and communicated the important stake the FRG has in the success of the Soviet Union's reform effort. Gorbachev also explained that it would take 20 years before *perestroika* could be successful. In May Gorbachev met for an hour with the head of the Free Democratic Party (FDP) and Minister of Finance (Martin Bangemann), articulating the important economic stake the West Germans have in a revitalized Soviet economy.

Gorbachev seems to be looking toward a new European phase in his detente policy toward the West. He has met with British Prime Minister Margaret Thatcher in London, French President François Mitterrand in Paris, and West German Chancellor Helmut Kohl in Moscow and Bonn. In this context, West European meetings could be used to address both economic and security issues, thus creating an "improved" environment for dealing with the Bush administration.

A key theme Gorbachev has articulated in his meetings with West German leaders and one firmly reinforced by Soviet representatives in the FRG is the importance of economic reform to the Soviet Union. The Soviets clearly perceive the West Germans to be key players in shaping Western policy toward East-West trade. They are trying to use various bilateral commissions as forums to convince West German businesspersons of the seriousness of change on the Soviet side. For example, at the first meeting of the West German-Soviet Commission for Scientific and Technological Cooperation, held on April 19, 1988, the agenda focused on questions of nuclear reactor safety, biotechnological research, and public health improvements. By focusing on issues seemingly outside the Cocom restriction area, the Soviets seek to defuse the political dimensions of East-West trade and to open a new chapter in West German-Soviet trade.

The loan for the construction of facilities to produce consumer goods in the Soviet Union contains an important innovation. The loan will be supported by issuing bonds that private citizens can purchase directly, creating for the first time the possibility of broadening participation in this bilateral relationship. Symbolic of the desire for broader scientific and technological cooperation has been the offer to allow a West German cosmonaut to participate in a Soviet space mission, probably at the end of 1992.

Parallel to the effort to reinvigorate the governmental relationship, the Soviets have sought to conduct a dialogue with the West German political parties as well. It is obvious that from the Soviet point of view the preferred party is the SPD. This is because the SPD is seen to pursue policies more compatible with Soviet definitions of their security interests in the West. It is interesting to note that only one staff member of the SPD *fraktion* in the *Bundestag* is really an Americanist; almost all others speak Slavic languages and travel to the East much more than to the United States.

After the change of government from the SPD to the CDU/CSU in 1982, the Soviet Union tried to strengthen its contacts with the new Kohl government and with the CDU. For example, Volker Ruhe (security policy spokesman of the CDU) visited Moscow in 1987 and met with high-ranking Soviet officials, including Zagladin. The most significant example of the changing Soviet attitude toward the CDU/CSU is unquestionably the invitation to Moscow of then Bavarian prime minister and chairman of the CSU, Franz Josef Strauss. Although Strauss has been, for the Soviets, the personification of revanchism, he was accorded all the honors of a state visit. Moreover, he was allowed to fly his private jet into the Soviet Union, becoming the first Westerner to do so. This deft touch left a profound impression on Strauss, and such flexibility is a key dimension of the new Soviet public diplomacy effort.

Members of the FDP, especially Foreign Minister Genscher, are viewed by the Soviets as the cofounders of West German *Ostpolitik*. Genscher's oft-repeated theme that it is in the Western interest to help Gorbachev succeed in his reform effort fits well with Gorbachev's basic message to the West, a message reinforced by Soviet representatives in the FRG. The message is this: "We are serious about economic reform. We will be heavily involved in domestic revitalization for a long time. We cannot be a serious threat to you if we are so heavily involved in domestic change, and it is in your interest to participate in our reform process."

The remaining political party, the Greens, is of limited interest to the Soviets, primarily because it is only a marginal party. The Soviets do have some contacts with the Greens, as evidenced in their invitation to two leading members of the party (Jutta Ditfurth and Jurgen Meier) to attend the celebrations of the 70th anniversary of the October revolution in Moscow. Also, other leading Green politicians (for example, Petra Kelly) had the opportunity to talk with Soviet Foreign Minister Eduard Shevardnadze during his visit to Bonn.

There may be one important lesson the Soviets (or at least some Soviets) have learned from the experiences of the Greens and the peace movement. The Greens were the most prominent political force in the FRG arguing that the deployment of the Pershing missiles in West Germany would dramatically increase the prospects for confrontation or war. Instead, the deployment led to disarmament, and the chief foreign policy spokesman for the Greens has publicly admitted that the party was acting on erroneous assumptions.

At least some Soviets understand that the Soviet military threat is a political factor in the West that worsens the security situation of the Soviet Union. For example, at a meeting among the SPD, a member of the Soviet General Staff, and a leading civilian proponent of reasonable sufficiency, open disagreement on the threat perception issue surfaced. The General Staff officer argued that the Soviet Union needed to rely on substantial conventional forces—especially tanks and artillery—to defend itself, whereas the civilian strategist argued that such forces are viewed as threatening in the West, regardless of whether the Soviet Union considered them to be defensive. Because the West did not consider these to be defensive weapons, it would seek to overarm in response. Therefore the Soviet Union must alter threat perceptions by a more effective disarmament policy.

In addition to government and party officials, the Soviets pay a great deal of attention to journalists. There are virtually no Soviet specialists among the West German journalists, who are generally more interested in changes in Soviet domestic policy than in foreign policy. The image conveyed by the West German press and by interviews with West German journalists is that the Soviet Union is in the throes of serious change. This focus reinforces the problem of conveying the continuing challenges that Soviet foreign and security policy presents to the West.

Illustrative of Soviet efforts to influence journalists more effectively is the marketing of their most important asset—Gorbachev—to West German journalists. For example, after the signing of the INF agreement, the Soviet embassy invited journalists in Bonn to an event similar to the U.S. government-sponsored Worldnet program. The journalists had the opportunity to see a televised speech by Gorbachev about the Washington summit and the INF treaty.

In addition, the Soviets have held their first meeting with the West German military. The Konrad Adenauer Foundation frequently organizes conferences in cooperation with the *Bundeswehr*. Soviet representatives attended one such conference and demonstrated a moderate tone in their statements and answers. They projected an image of being open-minded and willing to listen to criticism. Also, a leading West German expert on the Soviet Union with ties to the CDU has been invited by the Soviet embassy to give a lecture on the new Soviet military policy. This invitation is an important symbol of the new openness in Soviet discussions with West German elites.

In various meetings with West German security elites, Soviet representatives have indicated that they could accept the legitimacy of NATO, an American military presence in Europe, and the forward defense of West Germany. However, this requires considerably reduced force levels and a halt to, or at least control over, modernization. The Soviets are going out of their way to avoid the impression that they want to divide NATO or to separate the United States from Western Europe. For example, in mid-1988 the Soviet embassy declared its willingness to participate in a conference together with military representatives from the American and French embassies.

The Soviets are suggesting they are serious about changing their military doctrine as well. For instance, in March 1988 members of the general staff met with *Bundeswehr* officials. During this meeting the Soviets indicated that they were now elaborating the military-technical dimensions of the new security concept.

Soviet Public Diplomacy Themes: The Case of *Sowjetunion Heute*

One way to identify the content of Soviet public diplomacy themes in the FRG is to examine the content of the magazine designed to present the Soviet message to the West German public. The German language magazine *Sowjetunion Heute [Soviet Union Today]*, is published monthly by the public relations office of the Soviet embassy in West Germany in cooperation with Novosti. According to the magazine, its goals are to contribute to Soviet-West German relations and to the mutual understanding and cooperation between the two nations.[57]

Each issue of the magazine has one central theme, and reports are written in a good journalistic style and are presented in a modern manner. The journal provides documents on current Soviet policy as well. The magazine is provided free, with only a small postage payment requested. The articles since 1985 can be divided into three broad categories: those dealing with Soviet domestic affairs, events in the FRG, and foreign policy, especially disarmament issues.

Soviet Domestic Affairs

Coverage of domestic affairs focuses on propaganda regarding the quality of life in the Soviet Union; the diversity of the Soviet people; the development of Soviet science, technology, and economy; and elaboration of the reform process. The journal portrays an image of the Soviet Union as a society with important modern elements in the scientific and technological areas. In this context, Soviet success in space research is highlighted. The high standard of industrial production; the use of new production methods (including computers); successes in mechanical engineering, building, trade, and mining; and the achievements of Soviet shipbuilding and aircraft construction are noted as well.

Most important, the *perestroika* and new thinking themes are underscored. Until the sixth issue of 1987 the reform process merited only secondary attention. But since then there have been many articles written by a variety of authors, coming from various parts of Soviet political and cultural life, about the purposes and content of the new thinking. The authors try to persuade the readers of the bright future that lies ahead for the Soviet Union because of this change in its policy. The seriousness of Gorbachev's revolution (the change in state and society) is frequently repeated. Among the examples cited to prove this argument are the introduction of new working conditions (such as legitimization of small private businesses) and the new creativeness in art and culture. It is further argued that today it is increasingly possible to live a more individualistic life in the Soviet Union. For example, Western rock groups have appeared in the Soviet Union, and motorcycle fan clubs and even punks can be found in the streets of its cities.

Finally, improvements in the field of human rights are also noted. After a reform of the legal system, the individual now would be more protected against arbitrary acts of the state. Prominent Soviet specialists are cited as saying "You can breath easier now."[58] The image is one of progress under way despite the resistance to change.

Soviet-West German Relations

In the area of Soviet-West German relations, many key topics about city partnerships and meetings between groups from both sides are addressed in the "Contacts" column. For example, coverage was provided on the meeting of the FRG-Soviet working group held in June 1986 in Wiesbaden. In addition, there are reports about the possibilities for tourism in the Soviet Union. There are also discussions about economic relations on a case-by-case basis. Finally, special reports are printed on such occasions as the state visit of President Richard von Weizsaecker to the Soviet Union. Articles that examine the division of Germany after World War II make clear that it is not the Soviet Union that is to blame for this arrangement. With regard to the future of Germany and possible reunification, little hope is held out for resolving this issue.

Arms Control, Peace, and the United States

The main theme in these articles is the positive contribution the Soviet Union has made to the peace process. Various articles on arms control and arms reduction have argued that U.S.-Soviet negotiations have led to a reduction in the threat of war. The United States is portrayed as lagging behind the Soviet Union while, notably, the FRG's responsibility for helping advance the peace process is underscored.

Following the Washington summit, there was a noticeable shift in tone concerning the role of the United States. In contrast to earlier years, when the United States was accused of engaging in constant provocations, it was now argued that perceptions of the other side as an enemy should be reduced. In other words, the previous action-reaction cycle was also a threat to peace. The journal has focused little attention on Western intra-Alliance relations, in part to avoid the impression that the Soviet Union wishes to divide the partners of the Western Alliance from each other.

Key Images

The Soviets seek to convey four key images in this public relations journal. First, socialism is a desirable social order, and the effect of Gorbachev's reforms will be to enhance its desirability. Because of *glasnost'* and *perestroika*, the daily life of the Soviet citizen is becoming more and more comparable to Western standards. The central image is that the Soviet Union is becoming a normal, developed society. For example, one report, which discussed the daily life of people in the residential quarters of major cities, contained a picture of young women performing aerobics in modern outfits.[59] Only a short time ago such clothing was regarded as an example of Western decadence and would never have been featured in official publications.

The second major theme is the growing open-mindedness of the Soviet Union. The journal itself is trying to change its style so it can demonstrate this more liberal attitude toward modern and even unusual things. Third, the FRG-Soviet relationship is portrayed as a central one to the Soviet Union. There is always extensive coverage of the current bilateral relationship.

Finally, the editors seek to demonstrate the congruence of interests between the FRG and the Soviet Union. Emphasis is placed on Gorbachev's theme of a common European home in which only political arrangements and agreements can allow East and West to survive. The natural complementarity between the Soviet and West German economies is also recognized. German capabilities could be married to the vast Soviet natural resources to benefit both sides. Moreover, West German participation in the Soviet market reduces unemployment problems in the FRG. A growth in economic ties would lead to greater political predictability, which would be beneficial to West Germany. In other words, if West Germany pursues its real interests, cooperation with the Soviet Union will be at the center of German policy.[60]

West German Elite Reaction

Thus far, this chapter has identified West German reactions to the Soviet Union's public diplomacy only indirectly. Yet in fact much of the evidence of the Soviet public diplomacy effort has been drawn from interviews with West German elites who, at the same time, reveal their judgments of those efforts. Interviews with West German elites in the government, in the parties, and in the media reveal an intense interest in the scope and nature of change in the Soviet Union. After some initial skepticism, especially on the part of conservatives, virtually everyone acknowledges that a process of historic change is under way in the Soviet Union. There is considerable disagreement over how significant the changes have been to date as well as over how best to deal with the current situation.

Reaction to the Gorbachev phenomenon can be delineated into three broad groups. There are those who believe change is occurring but that it must be assessed, those who believe no change has occurred, and those who are enthusiastic about the transformation of the Soviet Union.

Wait and See

The largest group of elite opinion believes change is occurring but that the West should see whether Gorbachev's efforts can really make a difference. This group believes that it is necessary to see whether the new policy has a concrete chance of success, but it can be further divided into two subgroups over the question of how to respond to change. For some, it is a question of "engage and see," while for others it is a question of "observe and see."

Elites in this group believe it is advisable not to overestimate the Soviet Union's ability to change. But it is widely accepted that it would be desirable for the Soviet Union to succeed. The general sense is that for the country to succeed, it will have to turn its energies inward and become consumed with its own processes of change. It is widely believed that a turn inward, reform, and economic detente are closely interrelated.

These people consider it desirable for the FRG to expand its role in the Soviet economy. But interviews with West German businesspersons, bankers, and economic decision-makers reveal great skepticism concerning how quickly or effectively the Soviet economy can be changed. The West German businessperson is interested in making money from the Soviet market, whereas the government decision-makers intuitively perceive the key role that West German leadership of the West's effort to participate in the Soviet reform process could play in enhancing the role of the FRG both within the Alliance and in East-West relations.

The experience of the 1970s in dealing with the Soviets, as well as the Polish crisis in the early 1980s, have left their mark. West German economic elites are interested in expanded involvement in the Soviet economy but will insist on a detailed monitoring of changes in the Soviet foreign trade system. Government officials will undoubtedly dedicate a good deal of time during discussions with their Soviet counterparts to pushing the Soviets toward accepting the West German definition of necessary changes in Soviet foreign trade mechanisms.

For the first subgroup ("engage and see"), an active engagement policy with the Soviet Union is critical to maintaining public support for the FRG's current security concept. Gorbachev is perceived to have seriously altered the context within which Western security policy is made, and a passive "wait and see" policy would be disastrous. Instead, active exploration of the possibilities for change is required in order to determine the realities of change. This subgroup includes most economic elites, members of the foreign ministry, some members of the chancellory, some journalists, and some members of the CDU and FDP.

Most sectors of the German business community favor actively exploring the possibilities but are skeptical of rapid change. They would like to see a new line emerge in the economic relationship between the two countries, but they do not harbor many illusions about the problems to be encountered. As long as mere words fail to be supported by credible actions (such as placing important production facilities in the hands of real entrepreneurial managers), this group will remain skeptical. They are, however, quite interested in building stronger cooperation and in developing joint venture agreements. They will use joint venture talks to try to educate the Soviets about the requirements of contemporary markets.

For the second subgroup, an "observe and see" policy means doing nothing until "genuine" change occurs in Soviet policy. This subgroup represents the minority view within this first sector of elite opinion. This minority group includes members of the defense ministry, some members of the chancellory (and probably the chancellor himself), some journalists, and the more conservative members of the CDU. For this segment of the elite population, testing the seriousness of the Soviet reform effort might expose the West to new dangers: the cohesion within the FRG and the Alliance as a whole could be undercut.

No Change

The second group, a small minority of elite opinion, simply does not believe change has occurred at all. These are people whose image of the Soviet Union is influenced by their perceptions of the past and who are neither willing nor able to alter it. Their opinion is that the Soviets are really Russians, and Russians have never been efficient. The cultural heritage and future of Russia are to be governed by incompetent bureaucrats who will seek external expansion to rid themselves of domestic difficulties. The pretended reforms in the Soviet Union are, in reality, nothing more than an attempt to turn attention away from the real problem: the continued existence of communism.

One can find this opinion expressed especially on the far right wing of political life, for example, in refugee federations and some conservative student clubs. Many of the older generation of Sovietologists also belong to this group. But given the minimal influence of West German Sovietologists on elite opinion, their troglodyte attitudes are more self-serving than influential.

Enthusiasts

The third group consists of left-wing enthusiasts. For these persons, Gorbachev's very existence proves that their ideas about the desirability and possibility of significantly transforming the East-West system were simply ahead of their time. For them, Gorbachev is a virtual hero since he has raised the possibility of radically restructuring the military confrontation in Europe. For these elites, Western reluctance to work with Gorbachev is the problem, not any residual Soviet military threat.

This segment of opinion rests almost totally in the left wing of the SPD and in the Green party. It certainly is not a mainstream SPD position; the latter is much closer to Genscher's articulation about the necessity for active engagement.

The Parties and Reactions to Gorbachev

The CDU and CSU

Although there are some differences between the two sister parties, most leaders belong to the "wait and see" school of thought. For some in the CDU, active engagement is the preferable mode; others require a more passive observation mode. For most of the CSU, the observation mode is the most important, while some leaders continue to believe that no genuine change has occurred. The most conservative members of the CDU and CSU leaderships fall into the latter category, seeing Gorbachevism as only a significant new phase in the Soviet propaganda effort.

Volker Ruhe, the CDU spokesman in the *Bundestag* on security issues, illustrates the general CDU leadership position. According to Ruhe, the CDU would welcome a change in Soviet policy and a modernization of the Soviet Union because of Gorbachev's reforms.[61] Ruhe's trip to the Soviet Union in 1987 and his involvement in the difficult task of convincing West Germany's allies to resist including the Pershing IAs in the INF treaty have convinced him of the new climate in East-West relations. Ruhe believes in an active engagement policy incorporating pointed reminders to the public that many problems will remain in East-West relations as the restructuring process proceeds. As he noted in a May 1988 speech:

> Further development of the restructuring process in Soviet domestic and foreign policy is of major importance for the improvement of mutual security in the East and West. The opportunities that may derive from this should be made use of—soberly and to our mutual advantage. At the same time, we should not underestimate the dangers and difficulties involved

in the long-term process of political, social and economic restructuring. What and how much of past Soviet policies can and will be changed must be tested—carefully and uninfluenced by feelings of Euphoria—at the negotiating table and in practical cooperation. The yardstick applied should not be so much new thinking and new intentions—since both can change very rapidly—but rather new deeds on the part of the Soviet leadership.[62]

Ruhe's position reflects the mainstream position of the CDU and, in fact, he has played a major role in aggregating this position. His position supports much of Genscher's activity, but with much less expectation of success. He also reacted strongly to Genscher's notion that new thinking in the USSR should be met with new thinking in the West. For Ruhe, the Soviets are restructuring precisely because of the success of Western concepts, policies, and thinking; as such the West should not abandon its basic concepts but simply adapt them. As Ruhe has argued:

The realization on the part of the current Soviet leadership that the social and economic development of their country cannot go on as in the past, and that comprehensive restructuring is necessary in order to keep the gap that exists between the Soviet Union and "capitalist countries" from growing larger instead of smaller, is, in essence, a compliment to the West on the success of its policies. Mr. Gorbachev himself is the Soviet response to the success of Western policies and to the failure of Soviet policies in the decades that preceded his taking office.[63]

Ruhe has also been an important proponent of the need for a clear-cut comprehensive concept for security and disarmament (the so-called *Gesamtkonzept*) prior to any agreement to modernize NATO's remaining nuclear weapons deployed in Western Europe. The comprehensive concept aims for significant reductions in battlefield weapons and the establishment of a minimum deterrent level, even without negotiations. "The object of limiting the role of nuclear weapons in Europe to the absolute minimum in terms of both quality and quantity is to strengthen, not weaken, our strategy and Western defense. This may sound contradictory but it expresses our aim to maintain an acceptable and effective strategy of war prevention with markedly fewer nuclear weapons but with a more convincing structure."[64]

The chancellor's own public statements show a remarkable shift from stringent criticism of Gorbachev to much more hopeful expectation of Soviet change. The chancellor has moved from making analogies between Gorbachev and Goebbels to recognizing the serious possibilities of working with the Soviet Union. The change was evident in public positions he took in connection with the 1986 Reykjavik summit. He stated that this summit showed that the Soviet Union, and especially Gorbachev, is interested in an improvement in East-West relations.[65] Shortly thereafter, Kohl noted that Soviet foreign policy demonstrates new dynamics to which the FRG should pay attention.

Typical of Kohl's public commentary on the Gorbachev issue is a speech he delivered before the *Bundestag* in 1987. Since Gorbachev's advent to power, he has raised the people's hopes that the future will bring important changes in the Soviet domestic system. Kohl also pointed out that Gorbachev had spoken about "new thinking" in international relations, and he made clear that if this includes greater understanding, more cooperation, and especially concrete results in arms control, the West should respond in a positive manner.[67] Two months later, he noted more specifically that because Gorbachev's reforms pose the possibility for change in East-West relations, the FRG should play a more active role in exploring those opportunities.[68] Finally, in his speech on the INF treaty before the *Bundestag*, Kohl praised Gorbachev for removing some of the obstacles that had previously blocked the treaty's realization.[69]

The CSU leadership remains more skeptical, but even here movement is evident. After returning from his visit to Moscow in December 1987, Franz Josef Strauss noted that the Soviets were engaged in a sincere effort to change.[70] The May 1988 party conference in Munich was replete with statements that recognized the changes in the Soviet Union. For example, Hans Waigl, the leader of the CSU *fraktion* in the *Bundestag*, demanded that the West should not just react to the active policy of the Soviet Union but should make its own proposals for improving the East-West security situation. Strauss cautioned at the same conference, however, that one should not expect a quick transformation of the East-West situation, because the Soviet Union was still a communist system.[71]

The FDP

The FDP is clearly the most enthusiastic party in the government coalition with regard to the degree of Soviet change and the possibilities of working with the Soviet Union. The most visible spokesperson of the party is Foreign Minister Genscher. In a wide array of speeches, Genscher has underscored the importance of constructively engaging the Soviet Union in a process of change. Occasionally he uses such unfortunate phrases as "we should trust Gorbachev" or "we should modernize our thinking as well," but he does reflect a moving consensus in the West German elite about the significance of the changes in the Soviet Union.

Genscher has underscored that domestic change in the Soviet Union is serious. In connection with some highly critical remarks from politicians of the CDU/CSU right wing, Genscher warned that one should not dismiss developments in the Soviet Union as mere propaganda. It was an important task of the West German government to ensure that this did not happen.[72] He added later that the causes of the new Soviet policy are rooted in a domestic economic crisis and in the recognition that economic reform is the only way for the Soviet Union to compete internationally. The effort to carry out change in the Soviet Union is therefore serious.[73] More recently, Genscher has argued that even the process of democratization is becoming more serious in the Soviet Union as the *glasnost'* campaign incorporates a deeper process of social change.[74]

Other FDP leaders have been even more enthusiastic than Genscher. For example, State Minister Helmut Schafer sees evidence of new thinking in Soviet security policy. Soviet statements about the desire for "reasonable sufficiency" and "the necessity to think about a removal of certain asymmetries also in the conventional field" could well lead to a change even in the strategic doctrine of the Soviet Union with serious consequences for force structures and weapons acquisition.[75]

The SPD

The most influential politicians of the SPD view the Gorbachev reforms favorably and believe either in a policy of constructive engagement or belong to the enthusiast group. The most important SPD politicians who are involved in and determine their party's foreign and security policy are Willy Brandt (honorary chairman), Karsten Voigt (foreign policy spokesman of the SPD *fraktion* in the *Bundestag*), and Egon Bahr (chairman of the *Bundestag* arms control committee). They all perceive a serious process of change in the Soviet Union, are extremely sensitive to the visible moderation in the tone of Soviet policy, and see clear change in deeds as well.

Brandt is the most skeptical of these three leaders. Upon returning from his 1988 visit to Moscow, Brandt noted that although he had had a four-hour meeting with Gorbachev, he still did not understand what Gorbachev meant by the notion of a "common European home." He simply failed to find anything concrete in the term. Brandt did learn that Gorbachev believed that reform would be a long process, and Gorbachev conveyed in explicit terms the problems the Soviet leadership was having in implementing change. Gorbachev's frankness in discussing internal problems was unique in Brandt's dialogue to date with Soviet general secretaries.

Voigt sees encouraging signs for a new beginning in Soviet policy. The FRG can encourage change in Soviet foreign policy behavior by adopting a more open security concept rather than directly supporting democratization trends in the Soviet Union. He warns that it would be dangerous for the West to support reforms directly, for this would aid the conservatives in their arguments against reformers in the Soviet Union.

Bahr is by far the most enthusiastic of the SPD leaders. Because of Bahr's special efforts to negotiate a special German relationship in East-West relations with the East German Communist party, Bahr sees the Gorbachev reforms as a potential new opening in the ability of the two Germanies to work together. His views on the meaning of change in Soviet foreign and security policy are heavily shaped by this position. He has taken up the "common European home" theme with enthusiasm.

The Greens

The most enthusiastic of all West German elites about change in the Soviet Union clearly belong to the Green party. This party's position is well represented by a speech given by a foreign policy spokesman of the party, J. Meier, in Moscow in 1987. Meier argued that *glasnost'* and *perestroika* had altered the image of the Soviet Union in the West. He expressed the hope that *glasnost'* and *perestroika* would lead to democratization and the belief that many changes seemed to be profound. He then argued that the Soviet Union is no longer considered an enemy nation by the population of the FRG.[76]

Conclusions

The Soviets have developed a number of important public diplomacy themes during the Gorbachev period. Above all, they have promoted the ideas that there is now a new look in Soviet policy and that the nature of Soviet society is changing. The new thinking and *glasnost'* campaigns are being used to highlight this innovation in Soviet society.

Another major theme is that Western Europe and the Soviet Union share a common European home. Although the concept did not originate with Gorbachev, it has become a centerpiece of the current Soviet effort in Western Europe. The goal is to convince the West Europeans that they share many interests in common with the Soviet Union and that if a significant effort is made to overcome the old stereotypes, progress toward genuine interdependence is fully possible. In discussions with West Germans the Soviets emphasize the economic dimension of the common European home theme, especially in light of the widely shared belief within the West German elite of the legitimacy of expanding East-West trade.

In the wake of the INF agreement, the Soviets are stepping up their campaign for nuclear disarmament in Europe. The Soviets have clearly raised the hope that the West German government will support a third zero option in European nuclear arms control. This message was delivered during Shevardnadze's visit to Bonn but was met with little overt enthusiasm. The Soviets have established their position and will undoubtedly raise the public diplomacy stakes, if and when NATO decides to proceed with modernization of its short-range nuclear systems. The Soviets are carefully monitoring the arguments currently being made by West German elites regarding the problems posed by the short-range systems and will undoubtedly play back these arguments when the time is right.

West German policymakers are under serious pressure to try to maintain the current defense effort in the face of the effectiveness of current Soviet public diplomacy efforts. How does one take advantage of the opportunities proffered by the Gorbachev administration and at the same time make a serious effort to counter a continuing Soviet military threat? As the image of the Soviet threat recedes, the sacrifices required by maintaining a large conscript force will be increasingly difficult to justify. Certainly, West German policymakers believe that no increase in the West German defense effort is possible because of Gorbachev's success in modifying the threat perceptions of the West German public.

Above all, Soviet spokespersons are underscoring that there is a growing, more realistic possibility of East-West reconciliation if Soviet initiatives are met with correspondingly serious proposals by the West. Although the West is somewhat skeptical of the possibilities for Gorbachev's success in restructuring Soviet life, this message has been received and is being acted on. West German elites are closely following developments in the Soviet Union and are trying to encourage the success of the Gorbachev reform effort.

To the extent that Gorbachev can reform the Soviet economy to open up better prospects for success in East-West trade, the West Germans will be responsive. To the extent that the Soviets try to push security issues too quickly, the West German elites will be reluctant to move too far and too quickly. A careful balancing act on economic and security issues by Gorbachev will have an important impact on West German interest in and support for Soviet reform.

Notes

1. Research for this chapter was completed in August 1988.
2. I. Basova et al., "The FRG: Dangerous Tendencies," *MEMO*, 1 (1985), p. 59.
3. G. Kirillov, "Bonn: Peaceable Rhetoric and Militaristic Practices," *IA*, 4 (1986), p. 59.
4. Ibid.
5. Igor Borisov, "Bonn: Words and Deeds," *IA*, 10 (1987), pp. 41–42.
6. Ibid., p.42.
7. Ibid.
8. Ibid.
9. I. S. Kremer, *FRG: etapy vostochnoi politiki [The FRG: Stages of Its Ostpolitik]* (Moscow: Mezhdunarodnye otnosheniya, 1986), pp. 168–169.
10. Ibid., p. 181.
11. L. Bezymenskii, "Under a Canopy of American Missiles," *Pravda*, July 27, 1984, p. 4.
12. Basova et al., *MEMO*, 1 (1985), pp. 61–62.
13. G. Kirillov, "The Moscow Treaty: A Basis for USSR-FRG Relations," *IA*, 9 (1985), p. 32.
14. Ibid., pp. 34–35.
15. Kremer, *FRG*, pp. 184, 202.
16. Ibid., p. 204.
17. Interview with Yulii Kvitsinskii, as translated in *FBIS*, April 25, 1986, p. G2.
18. *FBIS*, July 21, 1986, p. G2.
19. Ye. Bovkun, "FRG: According to the Code of Good-Neighborliness," *Izvestiya*, August 11, 1986, p. 4.
20. *FBIS*, November 18, 1986, p. AA1.
21. "Dangerous Atavism," *Pravda*, November 27, 1986, p. 1.
22. *FBIS*, March 19, 1987, p. G6.
23. Yu. Yakhontov, "On the Debate in the Bundestag," *Pravda*, March 23, 1987.
24. Yevgenii Gulkin, as reported in *FBIS*, June 4, 1987, pp. H3–H4.
25. "M. S. Gorbachev's Meeting with R. von Weizsaecker," *Pravda*, July 8, 1987, p. 1.
26. Borisov, *IA*, 10 (1987), p. 47.
27. Ibid.
28. *FBIS*, December 30, 1987, p. 36.
29. A. N. Yakovlev, "The Dangerous Axis of American-West German Militarism," *SShA*, 7 (1985), p. 8.
30. Borisov, *IA*, 10 (1987), p. 41.
31. Yu. Yudanov, "Unemployment and the 'New Beggars' in the FRG," *MEMO*, 4 (1986), p. 114.
32. A. Demin, "A New Study on the Federal Republic of Germany," *IA*, 2 (1985), p. 141.
33. Interview with N. Ryzhkov, *New Times*, 38 (1986), as reprinted in *FBIS*, October 2, 1986, pp. R18–19.
34. L. Borisov, "The Federal Republic of Germany," *ZVO*, 5 (1985), p. 23.
35. E. Kondrashova, "The Militarism of West Germany: Dire Consequences," *IA*, 8 (1985), p. 87.
36. Kirillov, *IA*, 4 (1986), p. 55.
37. Yakovlev, *SShA*, 7 (1985), p. 9.
38. See, for example, Capt. 2nd Rank V. Kuzar', "A New Threat Is Developing," *KZ*, April 28, 1985, p. 3.
39. Ibid.
40. Kondrashova, *IA*, 8 (1985), p. 91.
41. Ibid., p. 89.

42 Basova et al., *MEMO*, 1 (1985), p. 57.
43 L. Bezymenskii, "The Wild Dreams of Modern Revanchists," *IA*, 3 (1985), p. 29.
44 Ibid., pp. 37–38.
45 A. Tsvetkov, "West German Intelligence Agencies Examined," *ZVO*, 9 (1986), as translated in JPRS-UMA-87-015, pp. 1, 8.
46 Kirillov, *IA*, 9 (1985), p. 33.
47 Kirillov, *IA*, 4 (1986), p. 60.
48 *FBIS*, March 16, 1987.
49 V. Bogachev, "Coveting the Nuclear Button?" *SWB*, May 5, 1987, pp. A1–2.
50 *FBIS*, August 10, 1987, p. AA3.
51 Borisov, *IA*, 10 (1987), p. 45.
52 Basova et al., *MEMO*, 1 (1985), p. 49.
53 Yakovlev, *SShA*, 7 (1985), p. 10.
54 A. A. Trynkov and S. Yu. Grigoryan, "Pro- and Anti-Americanism in the FRG," *SShA*, 6 (1987), p. 25.
55 See, for example, G. Zubkovin, *FBIS*, February 10, 1987, p. G1.
56 Trynkov and Grigoryan, *SShA*, 6 (1987), p. 33.
57 *Sowjetunion heute*, 9 (1986).
58 See, for example, *Sowjetunion heute*, 7 (1987).
59 *Sowjetunion heute*, 4 (1987).
60 *Sowjetunion heute*, 7 (1987).
61 *Deutschlandfunk*, March 20, 1988.
62 "Making Realistic Use of the Opportunities Offered by Restructuring," speech at Oxford University, May 23, 1988.
63 "Framing a Clear-Cut, Comprehensive Concept of Security and Disarmament," speech at Oxford University, May 24, 1988.
64 Ibid.
65 October 21, 1986, speech.
66 November 6, 1986, speech.
67 March 18, 1987, speech.
68 May 7, 1987, speech.
69 December 9, 1987, speech.
70 *Generalanzeiger* (Bonn), December 12, 1987.
71 May 7, 1988, speech at the foreign policy conference of the CSU held in Munich.
72 February 1, 1987, speech.
73 October 17, 1987, speech.
74 Speech before the European department of the Economic Advisory Board of the CDU in Bonn, May 6, 1988.
75 December 15, 1987, speech.
76 November 3, 1987, speech contained in *Verlautbarungen der Bundesgeschaftsstelle der Grunen*.

10
Soviet Challenges to NATO

Robbin F. Laird
Susan L. Clark

It is critical that the NATO Alliance effectively meet the challenges posed by the leadership of Mikhail Gorbachev. A key to this effort could be the emergence of a more mature partnership between the United States and Western Europe, which would present its own challenge to the Soviet Union. In fact, from Moscow's perspective, its primary concern is that three relatively distinct forms of Atlantic military power will in some way be cojoined: namely, that West Germany will become more nationalistic; the European NATO states will succeed, however imperfectly, in formulating a common military policy; and the United States will nurture both developments while at the same time maintaining a formidable military presence in the region. Soviet leaderships have tried to avoid abrupt shifts in both the tone and content of their West European diplomacy—shifts that the Atlantic countries might interpret as provocative or destabilizing. The cultivation of better bilateral relations, a grudging tolerance for the American presence on the European continent, and a strong preference for negotiation instead of confrontation are currently, and will probably remain, the favored mechanisms by which Soviet leaders will seek to arrest the further development of NATO Europe's military potential. In other words, the Soviets hope to contribute to the decline of Atlanticism without encouraging the further development of European cooperation in the economic, political, and military security areas.

Within this context, cohesion within NATO will obviously be of principal interest to Soviet leaders. They are aware of the varying degrees of cohesiveness among the NATO member countries; the West should assume that the Soviets will seek to test Alliance solidarity as much as possible, both in peacetime and during a crisis.

Alliance Cohesion

From the Soviet perspective, common interests among Alliance members are strong enough to preclude realistic expectations that the Alliance will dissolve in peacetime. Even if such dissolution occurred, a number of strong bilateral and multilateral arrangements among key Alliance members would likely remain intact. Nonetheless, within the Alliance there are various tensions, both between the United States and Western Europe and among the West European powers. These tensions provide important lines of fissure that can be played on in a crisis.

In other words, there is a general confluence of interest among the Western states, but tensions remain serious enough that different national interests persist that can be played on by the Soviet Union in attempting to shape Alliance policy. Cohesion exists, but it must be re-created in bargained relationships among the Alliance members. Those bargains can be called into question as specific national policies change or as pressures increase on the Alliance in periods of crisis.

The Soviets have identified a number of critical European-American fissures affecting Alliance cohesion. The Americans have global military reach and ambitions; European military power is predominantly regional in scope. The Americans have an interest in confining any East-West war that became nuclear to the European theater; the Europeans obviously do not. The Americans are interested in seeing the Alliance deploy forces for a protracted conventional war; the Europeans are postured solely for a short war option. The Americans tend to "militarize" relations with the Third World; the Europeans tend to emphasize political and economic power relations.

The Soviets have identified a number of pressure points in relations among the West European powers as well. The West Germans have a priority on intra-German cooperation, forward defense, conventional military operations, and the American nuclear guarantee, according to which the United States would escalate to global nuclear war if a European conflict occurred. The French oppose the development of too close an intra-German relationship, emphasize the defense of French territory, do not contribute to forward defense, emphasize nuclear independence at the expense of conventional operations, and have cast serious doubt on the ability of the Americans to guarantee West German security. The British place priority on the protection of British territorial integrity by deploying an independent nuclear force, deploying maritime forces in the North Atlantic, deploying forward forces in West Germany to initiate early nuclear fire in the event of war, and count on inextricably tying U.S. nuclear forces to Great Britain's defense through nuclear interdependence. In other words, there are conflicting national strategies among the major European powers that provide a potential pressure point for the Soviets in a major East-West crisis.

In assessing the security policies of individual West European powers, Soviet analysts pay a great deal of attention to political trends (in effect, the "political will" variable), the commitment of resources to defense (conventional versus nuclear, continental versus maritime), the orientation of the military forces (territorial defense, power projection), and the various bilateral and multilateral arrangements the country has established for the deployment and exercise of its forces. The Soviets are interested in both objective capability (the size and quality of forces) and subjective orientation (how those forces might be used to support allies as well as to provide for territorial defense).

Utilizing such an analysis, the Soviets might consider a neutral country, such as Switzerland, to be a potential ally of other European powers in time of war. Other countries are seen as current liabilities that could become greater assets if political changes occurred (such as Greece). Still others are viewed as dark horses with increasing military benefit (Italy and Spain). Put in other terms, Alliance cohesiveness is a dynamic process subject to change. A given country's contribution to Western defense will depend as much on political vitality as on the objective capability of its armed forces.

Implications for Western Policymaking

Above all, it is incumbent on Western policy analysts and policymakers to recognize the political strategy underlying Soviet policy toward the Western Alliance. Rather than focusing solely on weapon systems or military balances, it is necessary to address Soviet political strategy as well. As part of the West's dialogue with Gorbachev, it should insist on the recognition of the legitimacy of Western security interests and institutions. Western leaders should use summits and other so-called bully pulpits to convey to their publics that the cohesion of NATO and the legitimacy of such fundamental Western interests as the forward defense of West Germany and the continued presence of U.S. military forces must be recognized by Soviet leaders in order to make a serious East-West dialogue possible.

It is becoming increasingly necessary for senior U.S. policymakers to make a clear, explicit, and visible statement to the American people about why the United States has committed substantial forces to Western Europe. An integral component of the reaffirmation of the U.S. commitment to Europe must include visible and explicit encouragement for building the European pillar of the Western Alliance. In order to resist Soviet political blandishments toward Western Europe and to reduce the effectiveness of Soviet efforts to undercut the cohesion of the Western Alliance, it is absolutely indispensable that structural change be encouraged within the Alliance.

Associated with this structural change should be a much greater American recognition of and public support for the British and French nuclear deterrents. Whatever the military role of these forces, their existence is a key component of the overall Alliance effort to resist Soviet efforts to undercut nuclear deterrence within Europe. Soviet efforts to promote denuclearization within Western Europe primarily serve their political strategy of undercutting Alliance cohesion and of enhancing their influence within Western Europe; American policymakers should therefore recognize that arms control efforts must pursue political objectives as well as military ones.

In addition to underscoring the legitimacy of the political elements of the Western Alliance, it is necessary as well to argue for the minimum security levels that the West requires. Rather than offering disarmament as the ultimate goal of the East-West dialogue, the West should offer a much more realistic and desirable goal: ensuring a balance of military forces at a much lower level. The vision of minimum security requirements should include an insistence on the legitimacy of continuing many key elements of current Western policy but at lower levels of armaments. Above all, this means a continuation of nuclear deterrence in the form of a substantial nuclear presence by the United States in Western Europe as well as a continued substantial presence of American conventional forces in Europe.

The minimum security requirement concept could allow the Alliance to pursue policy on a unilateral basis rather than being held hostage to Soviet public diplomacy efforts alone. For example, the Alliance could decide to reduce the American battlefield nuclear component to 2,000 warheads (or some other level determined to be sound for military purposes) and declare that the West will go no further until there has been a fundamental restructuring of the conventional force balance. A minimum level of such weapons is necessary for deterrence. The Alliance would be stating, in effect, that there is no arms race in this category of weapons. Rather, it would challenge the Soviets to build down to NATO's lower level but not to eliminate such weapons altogether.

It is also necessary not to leave Gorbachev alone in offering a possibility for East-West reconciliation. Western efforts made in the 1950s could be revived, efforts that could incorporate the West's own vision of a resurgent Western Europe, free from Soviet domination. Perhaps the West can propose its own version of German reunification acceptable to the West, a Germany of genuinely free elections and an integral member of the European Economic Community.

In short, the thrust of the Gorbachev challenge lies in reopening the fundamental questions of the East-West relationship. The United States cannot meet this challenge by focusing on narrow and technocratic approaches to counting force balances and arguing the merits of some prospective conventional arms control treaty behind closed doors. Rather, we must offer our own statement of the political requirements for Western cohesion and East-West cooperation. To do less will be to leave Gorbachev in a position to manipulate the current phase of East-West relations decisively to Soviet advantage.

About the Authors

Robin F. Laird and Susan L. Clark are research staff members at the Institute for Defense Analyses in Alexandria, VA, where they have developed a policy program on Soviet and West European affairs. Their research specialties include Soviet security policy, Soviet foreign policy toward Western Europe and Japan, West European foreign and security policy, and arms control issues. Dr. Laird's most recent books include: *The Soviet Union, the West and Nuclear Arms* (New York: New York University Press, and London: Wheatsheaf Press, 1986); coeditor (with Erik Hoffmann), *Soviet Foreign Policy in a Changing World* (New York: Aldine, 1986); editor, *French Security Policy* (Boulder, CO: Westview Press, 1986); editor, *Soviet Foreign Policy* (New York: The Academy of Political Science, 1987); co-author (with David Robertson), *British Defense Policy* (Winchester, MA: Unwin Hyman, 1989); editor, *European Arms Control Policy* (Durham, NC: Duke University Press, 1989); and editor, *Strangers and Friends: The Franco-German Security Relationship* (New York: Simon and Schuster, 1989). Susan Clark's recent publications include (with Kenneth Maxwell), *Soviet Dilemmas in Latin America: Pragmatism or Ideology*, Critical Issues Series (New York: Council on Foreign Relations, September 1989); "Who Will Staff NATO?" *ORBIS*, Vol. 32, No. 4 (Fall 1988); "Japan's Role in Gorbachev's Agenda," *The Pacific Review*, Vol. 1, No. 3 (Fall 1988); and "The Demographic Challenge to NATO," *The Atlantic Community Quarterly*, Vol. 26, No. 1 (Spring 1988). She is the editor of *Gorbachev's Agenda: Changes in Soviet Domestic and Foreign Policy* (Boulder, CO: Westview Press, 1989).

Hannes Adomeit is associate professor of Soviet foreign policy and domestic politics at the Fletcher School of Law and Diplomacy, Tufts University, and was previously with the Stiftung Wissenschaft und Politik. He has published numerous articles on Soviet foreign and military affairs, and is the author of *Soviet Risk Taking and Crisis Behavior* (1982) and the co-author of *Foreign Policy Making in Communist Countries* (1979).

Charles Gati is professor of political science at Union College and a senior research fellow at the Research Institute on International Change, Columbia University. He is author or editor of seven books, including most recently *Soviet-East European Relations: An Introductory Essay* (1988) and *Hungary and the Soviet Bloc* (1986), which won the Marshall Shulman Prize of the American Association for the Advancement of Slavic Studies.

Erik P. Hoffmann is professor of political science at the Nelson A. Rockefeller College of Public Affairs, SUNY Albany, and a scholar-in-residence at the Institute for East-West Security Studies. He is the editor or co-editor of four anthologies on Soviet domestic international politics and the co-author (with Robbin Laird) of *Technocratic Socialism: The Soviet Union in the Advanced Industrial Era, "The Scientific-Technical Revolution" and Soviet Foreign Policy*, and *The Politics of Economic Modernization in the Soviet Union*.

Phillip A. Petersen is the Assistant for Europe and the Soviet Union on the Policy Support Programs Staff in the Office of the Deputy Under Secretary of Defense for Policy. He previously served as a research analyst for the Library of Congress and for the Defense Intelligence Agency. He has published numerous articles and books on Soviet defense policy and airpower.

Notra Trulock III is a senior fellow with the Strategic Capabilities Assessment Center of the National Defense University. His research concentrates on Soviet military theory and planning and military force capabilities. He is the author of several studies on Soviet perspectives on nuclear weapons, most recently *Soviet Perspectives on Limited Nuclear War*.

Index

Adamishin, Anatolii 58
Adenauer, Konrad 190, 191, 192
Afanasev, Viktor 82–3
Afghanistan 53, 198
  Soviet occupation of 65, 197
  U.S. involvement in 80
Africa 117
Air-land battle 201-2
Akhromeev, S. F. 18–19, 37
Andropov, Yurii 172, 198–9, 210, 217
Angola 80
Anticoalition strategy 1, 4–5, 6, 7, 8, 13, 16, 20, 44, 51–86, 163, 165, 177, 239, 252, 254
  effectiveness of 13, 56, 65–76, 81–2, 254
  and Gorbachev 16, 51–2, 54, 57, 60, 61, 62–4, 73, 77, 78, 81–2, 84–5, 102–3, 236, 251, 255
  objectives of 7, 51, 61–5
Antsiferov, Aleksandr 68
Anzherskii, S. 135, 136, 137
Arab-Israeli war 143, 197
Arbatov, Georgii 63, 80, 193
Arms Control (see also: Denuclearization; Intermediate-range nuclear forces; Nuclear-free zones) 4, 16, 26, 52, 58, 65, 66, 76–7, 85, 112, 113, 121, 130, 156, 157, 165, 166, 222, 223, 237, 238–9, 244, 254
  conventional 8, 17–18, 52, 58, 255
  European policy toward, role in 12, 67, 68, 71, 112, 121, 130, 132, 145–6, 151, 154, 156, 157, 224, 227–8, 229, 235, 238–9
  nuclear 8, 12, 17, 52, 58, 66, 69, 73, 77, 151, 247
  Soviet unilateral initiatives 17, 77
  verification 16
  zero options 16, 53, 67, 68, 163, 227–8, 247
Artyomov, V. 135–6, 141
Ascension Island 134
Atlanticism 2, 3, 9, 10, 71, 109, 118, 119, 120–1, 155, 230, 251
  British approach to 128, 130, 143, 153, 154
  French approach to 106, 107, 110, 111, 112, 118, 119, 120–1
  West German approach to 218
Ausland, John 168
Austria 89, 90, 95, 101, 187, 189

Backfire bomber 21
Bahr, Egon 231, 245, 246
Balance of power (see also: Correlation of forces) 2, 167, 168, 170
Balkans 95
Ballistic missile submarines (SSBNs)
  British and French 20–1, 118, 121, 138, 139
  U.S. 21, 110
Baltic
  military district 170
  Sea 167, 168, 169, 171–2, 173, 175, 200
  Soviet fleet 171, 177
  Straits 167
Bangemann, Martin 234
BAOR see British Army on the Rhine
Barents Sea 171, 174
Belgium 20, 44, 207
Bergedorfer Gesprächskreis 231
Beria, Lavrenti 189
Berlin (see also: Quadripartite Agreement on Berlin) 27, 233
  crises (blockade) 82, 182, 187, 188, 192
  wall 182, 190, 191, 192
Berlingske Tidende 173
Berner, Orlan 165
Biden, Joseph R. Jr. 210
Bilak, Vasil 89–90, 96
Biryusov, A. 134
Bogachev, Vladimir 67
Bol'shakov, V. 113
Borisov, L. 219
Bovin, Aleksandr 57–8, 59, 66, 70–1, 74–5, 80
Brandt, Willy 57, 182, 191, 193, 196, 219, 233, 245–6

Brezhnev doctrine 92, 193
Brezhnev, Leonid 52, 57, 93, 164, 170, 191
  and anti-coalition strategy 51–2, 62–3, 63–4, 73, 82
  foreign and security policy 76, 83, 92, 192–3, 194–6, 198–9, 210
Brezhnev period 16, 209
  stagnation 75, 209
British Army on the Rhine (BAOR) 137, 253
Brundtland, Gro Harlem 173
Brussels Commission (*see also:* European Community) 64
Bulgaria 94, 95, 96
Bundestag 196, 209, 235
Bundeswehr 6, 181, 190, 200–1, 219, 225–6, 236, 237
Burlatskii, Fyodor 61, 65
Bush, George 77, 234
Buskin, Yu. 111

Callaghan, James 130
Canada 11, 20, 148
Carlsson, Ingvar 176
Carter, Jimmy 130, 198
CDU, *see* Christian Democratic Union
Ceausescu, Nicolae 98
Central Committee of the CPSU 232, 233
Chad 117
Chemical weapons 72, 200, 219
Chernenko, Konstantin 198–9
Chernyshev, Vladimir 4
Chervov, N. F. 173, 204
Chevaline 138
China, People's Republic of (PRC) 121
Chirac, Jacques 67, 112, 113, 119, 151, 204
Christian Democratic Union (CDU) (*see also:* Federal Republic of Germany) 193, 195, 196, 199, 219–20, 222, 229, 231, 232, 235, 236, 241, 242–4, 245
  and Soviet Union 219–20, 222–3, 231, 232–3, 235
Christian Social Union (CSU) (*see also:* Federal Republic of Germany) 219–20, 222, 229, 235, 242–4, 245
  and Soviet Union 222–3, 235
CMEA *see* Council for Mutual Economic Assitance
COCOM *see* Coordinating Committee for Multilateral Export Controls
Cold war 74, 98, 99, 209
Committee for State Security (KGB) 37
Common European home 4, 7, 8, 17, 60–1, 68, 69, 70, 132, 230, 239, 245, 246, 247
Common Market *see* European Economic Community

Communist Party of Germany (KPD) 184, 195
Communist Party of the Soviet Union (CPSU) 30, 34, 37, 56, 57, 58, 72, 90, 182, 196, 199, 200, 227
Conference of the European Communist and Workers' Parties 194–5
Conference on Security and Cooperation in Europe (CSCE) 55, 69, 197
Conservative party (*see also:* Thatcher, Margaret; United Kingdom) 15, 128, 129, 130, 145, 151, 154, 156
  defense policy 130, 131, 145, 152, 153, 156
  domestic issues 152
  foreign policy 128, 129, 130, 154
  role in relations with Soviet Union 129, 130–3, 154
Conventional forces *see* Arms control; France; North Atlantic Treaty Organization; Soviet Union; United Kingdom; United States
Conventional war (option) 18, 21, 26, 28–9, 36, 40, 43, 121, 201, 252
Coordinating Committee for Multilateral Export Controls (COCOM) 52, 61, 234
Correlation of forces (*see also:* Balance of power)
  East-West 2, 5, 27, 34, 41, 42, 182, 188, 193
  U.S.–West European 119, 148
Council for Mutual Economic Assistance (CMEA) 85, 190, 196
  intra-ties 64, 75, 76
  relations with European Economic Community 52, 64, 71, 72, 233
CPSU *see* Communist Party of the Soviet Union
CSCE *see* Conference on Security and Cooperation in Europe
CSU *see* Christian Social Union
Cuba 197
Cuban missile crisis 183, 192
Czechoslovakia (*see also:* Prague Spring) 89–90, 94, 95, 96, 184, 196, 197, 201
  relations with Western Europe 89–90, 99, 100
  Warsaw Pact invasion of 92, 193

Davydov, V. F. 144, 147
Davydov, Yu. P. 109, 114
de Gaulle, Charles (*see also:* Gauliism) 105, 106–7, 108, 111, 116, 194, 195–6
de Llamby, General 111
Denisov, Yurii 166
Denmark 44, 176, 177, 205, 206
  Folketing committee 175
  and nuclear-free zone 165, 168, 169, 173, 174, 175, 176–7

Denuclearization, Soviet efforts (*see also:* Arms control; Nuclear-free zones) 8, 17, 26, 69, 163, 169, 171, 208, 218, 247, 254
Der Stern 209
Detente 8, 10, 17, 54, 56, 58, 65, 67, 80, 90, 94, 96, 97, 98, 99, 109, 128, 130, 156, 165, 166, 182, 193, 196–9, 218, 219, 220, 223, 225, 227, 229
DFU *see* German Peace Union
Ditfurth, Jutta 235
Djilas, Milovan 184
Dobrynin, Anatolii 83, 232, 233
Domestic reform, Soviet *see Perestroika*
Dukakis, Michael 77
Dulles, John Foster 190

Eastern Atlantic 133, 135, 136
Eastern Europe (*see also:* Bulgaria; Czechoslovakia; German Democratic Republic; Poland; Romania; Warsaw Pact) 4, 13, 59, 77, 89–103, 138, 139, 182, 190, 196, 203
 as factor in Soviet policy toward Western Europe 1, 13, 59, 90–1, 102
 Greek formula 102, 103
 reaction to INF deployment 96–8
 reform attempts 94
 relations with West 5, 58, 60, 63, 85, 90, 95–101, 230
 Soviet model in 89, 93–4
 and Soviet Union 13, 60, 63, 89, 91–4, 99, 100, 101–3, 182, 191, 193
East Germany *see* German Democratic Republic
EC *see* European Economic Community
Economic activity, Soviet
 joint ventures 16, 76, 241
 reforms *see Perestroika*
 relations with the West 16, 51, 52, 57, 60, 63–4, 65, 75, 77, 90, 94, 143, 197, 219, 220, 224, 225, 234, 239, 241, 247
Economic modernization, Soviet *see Perestroika*
EDC *see* European Defense Community
EEC *see* European Economic Community
EFA *see* European Fighter Aircraft
Eisenhower administration 27
Elysee Treaty 208
Emerging technologies 26, 33, 38, 41, 44, 45, 166
 blurring of offensive and defensive 41, 45
English Channel 133
Eureka 114
Eurogroup 11–12, 116, 129, 148, 149, 150, 205, 206, 218
Euromissiles *see* Intermediate-range nuclear forces
European Community *see* European Economic Community
European Defense Community (EDC) 189

European Economic Community (EEC or EC) (*see also:* West European economic cooperation) 2, 16, 64, 77, 82, 84, 107, 108, 114, 128, 143, 146, 147, 154, 190, 205, 207–8, 255
 relations with CMEA 52, 64, 71, 72, 233
 and Soviet Union 52, 62, 77
European Fighter Aircraft (EFA) (*see also:* West European defense cooperation) 136–7, 150, 151
Europeanization (*see also:* West European defense cooperation) 8–10, 11, 12, 13, 71, 118, 119, 120–1, 205, 230
 British approach to 127, 128, 130, 142, 146–51, 154, 155, 158
 French approach to 8–10, 107, 118–22
 West German approach to 218, 226
European nuclear force 10, 120, 121, 149, 204, 205, 207
European Parliament (*see also:* European Economic Community) 17, 64
European security cooperation *see* West European defense cooperation
European Union (*see also:* European Economic Community) 207–8
Eurostrategic war *see* Limited nuclear war
Evensen, Jens 174

Falin, Valentin 70–1, 80, 83, 84, 198, 209, 232, 233
Falklands conflict 134–5
FAR *see* Force d'Action Rapide
FDP *see* Free Democratic Party
FEBA *see* Forward edge of the battle area
Federal Republic of Germany (FRG) (*see also:* Arms control; Atlanticism; Christian Democratic Union; Christian Social Union; Europeanization; Free Democratic Party; Genscher, Hans-Dietrich; German Democratic Republic; German problem; Green Party; Kohl, Helmut; Nazis; Nuclear deterrence; Nuclear hostage; Peace movement; Social Democratic Party; West European defense cooperation) 5–8, 20, 21, 27, 77, 90, 102, 136, 137, 147, 150, 181–211, 217–47, 255
 CDU/CSU government 15, 219–20, 222, 229
 CDU/SPD government 193–4
 defense budget 224, 230, 247
 doctrine and military strategy 6–7, 200, 201–3, 225
 domestic issues 182, 189
 economy 101, 188, 223–5
 foreign policy 81, 182, 195, 197, 218–19, 220, 245
 military capability 20, 182, 187–91, 200, 205–6, 211, 218, 225–6, 247

Federal Republic of Germany *(continued)*
  nuclear aspirations 10, 79, 120, 195, 203–5, 225, 226, 228
  ostpolitik 182, 193, 194, 195, 196, 220, 229
  reaction to changes in Soviet Union 239–46
  relations with the United States 5–6, 72, 79, 119, 143, 194, 196, 199, 219, 220, 225–6, 229, 230, 231
  revanchism 20, 72, 89, 189, 194, 195, 198, 200–1, 206, 218, 222, 223, 226, 227, 230, 233, 235
  role in NATO Western Europe 6, 12, 13, 79, 114, 119, 120, 149, 182, 190, 194, 195–6, 200, 206, 218, 223, 225, 226–9, 240, 252–3
  security policy *(see also:* Forward defense) 6, 110, 132, 181–3, 196–9, 226, 230, 241, 245, 251, 252
  and Soviet Union, Eastern bloc *(see also:* Moscow Treaty; Nuclear hostage; West German–Soviet Commission for Scientific and Technological Cooperation) 5–8, 15, 52, 53, 57, 62, 66, 72, 101, 102, 181–211, 217–47, 252
  Soviet wartime strategy toward 20–1, 27
  SPD/FDP government 196–9
Feindbild 209
Fencer 21
Finland 91, 95, 169, 189
  and nuclear-free zone 164, 165–8, 171, 173, 176
Flexible response strategy 7, 28, 30, 33, 168, 201, 202
FOFA *see* Follow-on forces attack
Follow-on forces attack (FOFA) 202
Force d'Action Rapide (FAR) 20, 111
Foreign Affairs Commission 83
Foreign policy, Soviet *see:* Brezhnev, Leonid; Eastern Europe; Federal Republic of Germany; France; Gorbachev, Mikhail; New thinking; Soviet Union; United Kingdom; United States
Forward defense 6, 201–2, 236, 252, 253
Forward edge of the battle area (FEBA) 43, 44
France *(see also:* Arms control; Atlanticism; Chirac, Jacques; de Gaulle, Charles; Eureka; Europeanization; Gaullism; Mitterrand, François; Nuclear deterrence; West European defense cooperation 8–10, 20, 27, 98, 105–22, 132, 147, 150, 151, 205, 206, 223–4
  anti-Americanism 106, 121
  arms sales 107, 116, 120, 122
  as a world power 107, 117
  cohabitation government 112, 113, 218
  conventional forces 105, 115, 116, 117
  domestic politics 109, 112, 113
  force posture 111, 114–18
  independence 10, 105, 106, 110, 111, 112, 114, 115, 116, 117, 119, 120, 122, 252
  Military Program Law 108, 110
  military strategy and doctrine 9, 108, 115, 116, 252
  nuclear weapons *(see also:* Ballistic missile submarines) 8, 9, 20, 21, 67, 68, 69, 105, 106, 107, 108, 110, 111, 112, 114, 115–16, 117, 119, 121, 122, 132, 156, 203, 204, 205, 207, 228, 252, 254
  out-of-area 3, 111, 117, 121
  relations with United States 106–7, 108–9, 110, 115, 117, 144
  role in NATO, Western Europe 13, 105, 106, 108–9, 110, 111, 114, 115, 116, 117, 118, 119, 120, 206
  security policy 105–22, 218
  and Soviet Union 8–10, 66, 92, 107, 109, 110, 111–12, 113, 114, 192, 194, 196
  Soviet wartime strategy toward 20–1, 105–6, 117–18
  withdrawal from NATO military organization 106, 118, 122, 141, 194
Free Democratic Party (FDP) *(see also:* Federal Republic of Germany; Genscher, Hans-Dietrich) 195, 196, 199, 219, 229, 234, 235, 241, 244–5
  and Soviet Union 195
FRG *see* Federal Republic of Germany
Frog artillery rocket 192

Gaivoronskii, F. 43
Galkin, Yu. 135
Gareev, M. A. 19, 31, 32, 36, 38
Gas pipeline (Soviet-West European) 52, 111, 143, 156
GATT *see* General Agreement on Tariffs and Trade
Gaullism 10, 106–7, 118, 121
GDR *see* German Democratic Republic
General Agreement on Tariffs and Trade (GATT) 77
General Staff 31, 32, 38, 204, 235, 237
  Military Science Directorate 38
General Staff Academy *see* Voroshilov General Staff Academy
Genscher, Hans-Dietrich 220, 221, 222, 227, 235, 242, 243, 244–5
Gerasimov, Gennadii 83–4, 145
German Democratic Republic (GDR) *(see also:* Eastern Europe; Peace movement; *Westpolitik*) 94, 95, 96–7, 101–2, 182, 183, 190, 191, 194, 195, 201, 210

German Democratic Republic *(continued)*
　relations with the Federal Republic of Germany 96, 97, 194, 197
　relations with the Soviet Union 97
　relations with Western Europe 90, 99, 100, 101, 246
German, Robert 177
German Peace Union (DFU) 195
German problem 183–7, 190, 191–200, 209–11, 238
Gesamtkonzept 243
Giscard d'Estaing, Valéry 106, 108–9, 111
Glasnost' 15, 17, 56, 90, 181, 232, 233, 239, 245, 246
Golubev, A. V. 130, 145, 156
Gorbachev, Mikhail (*see also:* Anti-coalition strategy; Murmansk speech; New thinking; *Perestroika*; Public diplomacy; Reykjavik summit; Soviet Union; United Nations) 4, 5–6, 15, 52, 55, 56, 57, 66, 67, 68, 70, 79, 86, 94, 111, 131, 155, 157, 182, 209, 218, 221, 222, 223, 232, 233, 235, 238, 242, 243, 244, 245, 246, 247, 255
　changes in foreign and security policy 14–18, 27, 30, 56, 75–6, 80, 83, 91, 92–3, 243, 244, 246, 255
　and ideology 4, 60
　image 17, 85, 236
　peace offensive 53, 66
　personality and style of 13, 14, 17, 55, 78, 93, 132, 164, 177, 236, 245–6
　speeches 17, 58, 82, 89, 172–4, 199–200
　(use of) summits 14, 15, 52, 54, 66, 67, 77, 111, 112, 113, 131, 132, 145, 147, 176, 177, 222, 223, 233–4, 236, 244, 245
Gorshkov, Nikolai 152
Great Britain *see* United Kingdom
Great Patriotic War 29
Grechin, S. 135, 139
Greece 91, 95, 102, 205, 206, 253
Greenland Sea 173
Green party (*see also:* Peace movement Federal Republic of Germany) 235, 242, 246
Grenada 53, 80
Griffiths, Franklyn 82
Gromyko, Andrei 14, 15, 55, 58, 210, 217
Gusenkov, V. 79

Hades missile 204
Healy, Denis 155
Heath, Edward 128, 154
Hernu, Charles 204
Heseltine, Michael 130, 154
High Commands of Forces 41
Holy Loch 129

Honecker, Erich 96–7
Horizontal escalation 42, 198, 202–3
Howe, Geoffrey 130, 131, 132–3, 145
Human rights 52, 90, 112, 130, 238
Hungary 94, 95, 96, 98, 101, 103, 184, 190
　and Western Europe 98, 99, 100, 101
　Soviet invasion of 92, 190

Iberian Basin 135
ICBM *see* Intercontinental ballistic missile
Iceland 168
IEPG *see* Independent European Program Group
IMF *see* International Monetary Fund
Independent European Program Group (IEPG) (*see also:* West European defense cooperation) 11, 146, 150, 205, 206, 218
INF *see* Intermediate-range nuclear forces
Institute of World Economies and International Relations 83
Institutniki 15
Intelligentsia 15, 16
Intercontinental ballistic missile (ICBM) 27, 28, 139, 169, 192
Intermediate-range nuclear forces (INF)
　British and French inclusion/exclusion in negotiations 112, 114, 115, 121, 132, 156
　compensation for 121, 146, 151, 164, 173
　NATO deployment of 6, 7, 18, 37, 53, 54, 76, 80, 85, 96, 98, 110, 115, 153, 164, 165, 166, 168–9, 182, 197, 198, 199, 202, 203, 206, 210, 217–18, 220, 223, 225, 226–9, 230, 235
　negotiations 66, 67, 68, 86, 96, 97, 112, 114, 115, 121, 132, 156, 228
　Pershing-1As 228, 232, 242
　Soviet attitude toward 6, 7, 15, 37, 57, 76–81, 72
　Soviet forces 6, 7, 27, 132, 192, 199
　treaty 4, 7, 17, 77, 82, 121, 132, 145, 146, 151, 156, 164, 169, 173, 177, 229, 236, 242, 244, 247
　Western German role in deployment decision 226–9, 230
International Department 61, 83, 232, 233
International Monetary Fund (IMF) 77
Ionov, M. 29
Ireland 205
Israel 108
Italy 3, 12, 27, 92, 119, 149, 150, 253
Ivanov, A. 141–2
Izvestiya 27, 57, 67, 140, 151

Jaguar 136–7, 149–50, 204
Japan 58, 64, 70, 75, 77, 78, 81, 101, 148
Johnson, A. Ross 85

Kadar, Janos 98
Kalachev, M. 78
Karemov, A. 115–16, 133–4
Kekkonen, Urho Kaleva 164, 166
Kelin, V. 153
Kelly, Petra 235
Kenney, Brian 188
KGB *see* Committee for State Security
Khesin, E. S. 11, 128–9, 130, 148
Khrushchev, Nikita 27, 34, 83, 94, 190, 191, 192
 and anti-coalition strategy 51–2, 62–3, 82
 post- period 35
Kiesinger, Kurt-Georg 193
Kinnock, Neil (*see also:* Labour party) 131, 153, 156
Kirillov, G. 220
Kir'yan, M. M. 41
Kohl Helmut 57, 198, 219, 220, 221, 223, 225, 229, 234, 235, 243, 244
 Goebbels statement 221–2, 243
 and Reykjavik 67, 227, 244
Koivisto, Mauno 172–3, 176
Kola Peninsula 170, 172, 175, 177
Koloskov, I. A. 115
Kolosov, G. V. 11–12, 139, 140, 141, 142, 144, 145, 146–7, 148–9, 150, 154
Komissarov, Yurii 167, 168, 170, 172, 175
Kommunist 37
Komsomol 201
Kondrashev, Stanislav 67
Konrad Adenauer Foundation 236
Korean War 187, 188
Kostikov, M. 169
Kosygin, Alexei 167
Kovalev, A. 108
KPD *see* Communist Party of Germany
Krasnaya zvezda 73, 169
Kudryavtsev, A. 109–10, 113–14
Kulikov, V. G. 37
Kurier 58
Kuzar', V. 121, 226
Kvitsinskii, Yulii 209, 221, 231–2

Labour party (*see also:* Kinnock, Neil; United Kingdom) 67, 128, 129, 151, 153, 154, 155, 156
 defense policy 128, 129–30, 131, 152, 153, 155
 foreign policy 128, 129
 fractionalization 131, 152
 role in relations with Soviet union 129–30, 131, 153, 155
 unilateral nuclear disarmament 131, 152, 153
Laird, Robbin 81
Leander frigate 135
Lebanon 117, 121

Le Monde 79
Lenin, V. I. 83
Leningrad military district 170
Leskov, V. 141
Libération 79
Libya, U.S. strike against 53, 80, 143, 155
Ligachev, Yegor 56, 83, 112, 170–1, 172
Limited nuclear war option
 Soviet ability to conduct 7, 19, 40
 U.S. NATO doctrine of 7, 9, 21, 29, 30, 38–9, 119, 133–4
 West European reaction to 9, 118, 133–4, 140
Lomeiko, Vladimir 55
Losev, Sergei 232
Lukin, V. P. 118, 119–20
Lyutyi, Aleksandr 146, 151

Madzoevskii, S. P. 11, 128–9, 130, 148
Malenkov, Georgii 189
Marov, Yu. 134
Marshall Plan 185
Marxism-Leninism 182, 184, 185
Maslennikov, Arkadii 130–1, 152
Massive retaliation 28, 29
Matsulenko, V. 19
MBFR *see* Mutual and Balanced Force Reductions
Mediterranean Sea 135, 172
Meier, Jurgen 235, 246
Mel'nikov, D. E. 197
Méry, Guy 108
Middle East 107–8, 117, 122, 143
Mikolajczyk, Stanislaw 184
Military doctrine, Soviet (*see also:* Reasonable sufficiency) 17, 18, 26, 36–7, 237, 245
 defensive orientation 6, 29, 31, 32
Military strategy, Soviet (*see also:* Conventional option; Limited nuclear war option; Nuclear war; Nuclear weapons; Theater warfare) 19–20, 21, 26, 33, 36, 46, 121–2, 236
 coalition 18–19, 121–2
 political-military aspect (*see also:* Anti-coalition strategy) 1, 2, 6, 18–22, 32, 253
 strategic objectives 31–4, 40–4
 technical-military aspect 2, 18–22, 25–46, 253
 Western view of 30, 236
Military thought *see Voennaya mysl'*
Ministry of Foreign Affairs 14, 233
 changes under Gorbachev 14–15
 Information Department 83
Mirage aircraft 204
Mirovaya ekonomika i mezhdunarodnye otnosheniya 167
MIRV *see* Multiple independently targetable reentry vehicle

Index

Mitterrand, François 109, 112, 113
  and foreign and security policy 10, 106, 109–14, 119
  relations with the Soviet Union 112, 113, 234
Molotov, Vyacheslav 185
Mongolia 77
Moscow News 233
Moscow Treaty 197, 220
MRV *see* Multiple reentry vehicle
Multinational corporations 77
Multiple independently targetable reentry vehicle (MIRV) 138
Multiple reentry vehicle (MRV) 138
Murmansk speech 172–4, 176, 177
Mutual and Balanced Force Reductions (MBFR) 197

Nationaldemokratische Partei Deutschlands (NPD) 195
NATO *see* North Atlantic Treaty Organization
Nazis 6, 19, 184, 195, 201, 211
Neisse River 185, 187
Netherlands 20, 44, 174
Neues Deutschland 97
Neutron weapons 166, 203, 204
New Statesman 146
Newsweek 188
New thinking (*see also:* Gorbachev, Mikhail) 16–17, 30, 55, 82–3, 171, 181, 209–11, 221, 232, 233, 237–8, 243, 244, 246
New York Times 27
Nicaragua 80
Nikolaev, N. 138–9, 145–6
Nimrod 150
Nixon, Richard 57
Nonprovocative defense 18, 231
North Atlantic Treaty Organization (NATO) (*see also:* individual member country listings, as well as: Air-land battle; Atlanticism; Eurogroup; Europeanism; Flexible response strategy; Follow-on forces attack; Massive retaliation) 16, 17, 30, 42, 44, 52, 77, 106, 107, 111, 118, 129, 136, 140–2, 148, 149, 153, 158, 164, 165, 169, 188, 189, 190, 203, 205, 206, 230, 251
  coalition strategy 12, 141, 149, 201
  cohesion of 19, 27, 29, 32, 52, 54, 58, 66, 82, 84, 85, 122, 163, 165, 174, 177, 230, 241, 252, 253, 254, 255
  conventional forces 6, 10, 12, 38, 117, 157, 200, 202, 225, 226, 254
  differing U.S.–West European interests 4, 5, 6–7, 9, 27, 53, 60, 69, 70, 239, 252
  European pillar 8, 11, 81, 111, 114, 121, 146, 155, 218, 230, 254

mobilization potential 19, 20
nuclear forces (*see also:* France; Nuclear weapons; United Kingdom; United States) 17, 21, 26, 30, 34, 76, 105, 110, 115, 116, 119, 139, 145, 151, 158, 163, 225, 232, 243, 247, 254
nuclear strategy (*see also:* Limited nuclear war option; Nuclear war; Theater warfare) 11, 28–9, 36, 144, 177
out-of-area (*see also:* France; United Kingdom; United States) 3, 7, 10, 117, 121, 252
Soviet peacetime strategy toward 4–5, 16, 26–31, 52, 236, 252, 254
Soviet wartime (crisis) strategy toward (*see also:* Military strategy) 18–22, 25–6, 30, 31–44, 252
ties with Warsaw Pact 71, 73
U.S.–West European security relations 2, 3, 4, 6–7, 9, 10, 17, 51, 54, 77, 81, 108, 111, 114, 119, 146, 158, 205, 207, 218, 230, 251, 253–5
North Sea 135, 171, 173
Norway 174, 177, 205
  Colding report 175
  and nuclear-free zone 165, 168, 169, 171, 173, 174–5, 176–7
Norwegian Sea 169, 171, 173, 174
Novoe vremya 68
NPD *see Nationaldemokratische Partei Deutschlands*
Nuclear deterrence 7, 19, 25, 27, 28, 38, 145, 157, 168, 177, 204, 254
  British attitude toward 6, 11, 132, 133, 145, 146, 157
  French attitude toward 6, 115, 132
  West German attitude toward 6, 132
Nuclear escalation 19, 21, 26, 30, 36, 37, 38, 119, 122, 140
Nuclear-free zones (*see also:* Denuclearization) 60, 131
  in Northern Europe 163–7
  Soviet initiatives (*see also:* Murmansk speech) 168–74
  Soviet support for 163–8
Nuclear hostage
  and Federal Republic of Germany 5, 6
  and Western Europe 27, 72, 191, 199
Nuclear war (*see also:* Limited nuclear war option; Theater warfare) 9, 19, 25, 28–9, 34–5, 36, 61, 82, 119, 133, 140, 155, 192, 193, 201, 227, 252
Nuclear weapons (*see also:* Arms control; Ballistic missile submarine; Denuclearization; European nuclear force; Federal Republic of Germany; France; Intercontinental ballistic

Nuclear weapons *(continued)*
  missile; Intermediate-range nuclear forces; Military strategy; North Atlantic Treaty Organization; Nuclear-free zones; Soviet Union; Treaty on the Non-Proliferation of Nuclear Weapons; United States; United Kingdom) 6–8, 12, 17–22, 25–41, 72–3, 76–81, 96–8, 105–22, 130–42, 144–6, 151–7, 163–77, 191–3, 201–5, 226–8, 247, 252–3
  dividing nuclear from nonnuclear powers 8, 163, 165, 177
  European nuclear capability 10, 12, 20, 21, 67, 68, 69, 81, 151
  first use 34, 202
  nonproliferation 167, 168
  parity 7, 25, 27, 170
  pronuclearism 4, 11, 67–8, 144, 145, 154, 156
  role in theater warfare 34–40, 140
  use of 19–20, 21, 25–41, 202, 211

Oder river 187
Ogarkov, N. V. 29, 36, 38, 116
OMGs *see* Operational mobile groups
"On Reorganization and the Party's Personnel Policy" 89
Operational mobile groups (OMGs) 43, 45
Owen, David 63

Palme, Olaf 176
Peaceful coexistence 57, 82, 83, 84, 90, 219
Peace movement 15, 53, 164, 167, 182, 199–200
  in the Federal Republic of Germany 5, 182, 199–200, 217, 230, 235
  in the German Democratic Republic 97
  in the United Kingdom 131, 153
Penkovsky, Oleg 190–1
Perestroika 14, 90, 132, 133, 231, 232, 233, 234, 237–8, 239, 246, 247
  domestic reform 14, 15, 52, 56, 157, 233, 235
  ecomomic modernization and reform 14, 51, 75–6, 77, 224, 234, 235, 240, 245
Perestroika: New Thinking for Our Country and the World 17
Perov, G. 138, 139
Pershing missiles *see* Intermediate-range nuclear force
Persian Gulf 121, 198
Pluton 204
Podgornii, N. V. 166
Poland 41, 64, 80, 94, 95, 96, 101, 103, 184, 190
  1980–81 crisis 53, 65, 92, 197, 240
  postwar settlement 185, 186, 187
Polaris 135, 138, 153
Politburo 52, 54, 57, 58, 60, 67, 77, 85
Pompidou, Georges 107–8

Ponomarev, Leonid 78–9
Popescu, Dumitru 99
Portugal 205
Portugalov, Nikolai 209, 232
Power center, West European 3, 106, 108, 109, 118, 119, 120, 148
  military 3, 9, 10, 109, 118, 119, 120, 148
  political-economic 2, 108, 119, 224
Prague Spring 92, 94, 196
Pravda 64, 72, 82, 97, 99, 113, 116, 146, 152, 169, 210, 221
PRC *see* China, People's Republic of
Primakov, Yevgenii 83, 84
Prussia 185
Public diplomacy 1, 14–18, 37, 77, 157, 163, 217, 251, 254
  interdependency 17, 18
  under Gorbachev 230–47

Quadripartite Agreement on Berlin (*see also:* Berlin) 197

Rapid Action Force *see* Force d'Action Rapide
Reagan, Ronald 4, 37, 57, 67, 68, 77, 83, 84, 85–6, 132, 145, 152, 229
  foreign policy 72, 130, 198
  military buildup 65, 76
Reasonable sufficiency 18, 231, 235, 245
  connection with nonprovocative defense 18, 231
Reform *see Perestroika*
Reykjavik summit 4, 53, 67–70, 71, 84, 145, 151, 156, 170, 171, 218, 222, 227, 244
  West European reaction to (*see also:* Kohl, Helmut; Thatcher, Margaret) 12, 67–8, 132, 145, 151
Rogers Plan 201–2
Romania 41, 94, 95, 96, 97–8
  and Western Europe 99, 100, 101
Roshchupkin, V. 155
Rotmistrov, P. A. 27
Rude Pravo 89, 99
Ruhe, Volker 235, 242–3
Ruhl, Lothar 231
Ryzhikov, V. A. 129–30, 153
Ryzhkov, Nikolai 173, 224

SALT *see* Strategic Arms Limitation Talks
Saudi Arabia 150
Savanin, L. 169
Scandinavia 163, 166, 169, 172, 174–6
Schafer, Helmut 245
Schmidt, Helmut 198, 199, 219, 231
Scud missile 191
SDI *see* Strategic Defense Initiative

SED *see* Socialist Unity Party
Semin, G. 115–16, 133–4
Semyonov, Vladimir 232
Serebryannikov, A. 74
Shavrov, I. E. 28, 31, 39
Shevardnadze, Eduard 14, 68–9, 131, 132, 221, 228, 235, 247
Shiryaev, P. 136
Shishlin, Nikolai 69–70
Slavenov, V. P. 108
SLBM *see* Submarine-launched ballistic missile
Social Democratic Party (SDP) (*see also:* Federal Republic of Germany) 15, 67, 184, 195, 196, 199, 219, 229, 230, 231, 233, 235
  Godesberg party program 195
  and Soviet Union 195, 218, 230, 231, 234–5, 242, 245–6
Social Democratic Party (SDP)–Liberal alliance (UK) 151, 153–4
Socialist Unity Party (SED) 195
South Africa 80
Soviet Military Encyclopedia 36
Soviet Union (*see also:* Anti-coalition strategy; Arms control; Economic activity; Eastern Europe; Federal Republic of Germany; France; Gorbachev, Mikhail; Military doctrine; Military strategy; Peaceful coexistence; United Kingdom; United States; Warsaw Pact)
  arms sales 120
  conventional forces 17, 27, 28, 31, 34, 38, 68, 76, 86, 175, 199, 236
  conventional superiority 175, 191, 228, 235–6, 245, 254
  domestic policy (*see also:* Perestroika) 236, 237–8, 243
  military power 191–3, 209
  navy 163, 167, 172, 174
  nuclear weapons 21, 27, 28, 34, 55, 155, 156, 170, 172, 173, 175, 177, 188, 191, 192, 199
  role of embassies 15, 232–3, 236, 237
  security 183–7, 188, 206
  training and exercises 29–30, 31
  utility of military power 26–7, 30–1, 182, 192–3, 198, 219
Sowjetunion Heute 237–9
Spain 9, 119, 253
SPD *see* Social Democratic Party
Spiegel 70–1, 84
SSBN *see* Ballistic missile submarine
Stalin, Joseph 13, 51, 83, 91, 92, 93, 183, 184, 185, 187, 189, 191, 192
  postwar policy 183, 185–9
  post- period 60, 94
State Treaty of 1955 187

Stent, Angela 62
Strategic Arms Limitation Talks (SALT) 52
  ratification abandoned 66, 197
Strategic Defense Initiative 52, 53, 54, 57, 84, 85, 86, 113–14, 220, 225
  Soviet reaction to 76–81, 113
Strauss, Franz Joseph 15, 223, 233, 235, 244
Submarine-launched ballistic missile (SLBM) 138
Suez crisis 192
"Suez syndrome" 135
Sulitskaya, T. I. 108
Sweden 166, 168, 169, 174, 176
Swedish Institute of International Affairs 166
Switzerland 253
Szuros, Matyas 98–9

Tamanskii, V. 142
Tarsadalmi Szemle 98
Tehran conference 185
Teltschik, Horst 231
Thatcher, Margaret (*see also:* Thatcher, Margaret; United Kingdom) 128, 131, 132, 143, 151, 152, 155–6
  foreign and security policy 130, 145, 152, 154
  influence on U.S. policy 12, 15, 145, 151
  relations with the Soviet Union 130, 131–2, 234
  and Reykjavik 12, 67, 68, 132, 145, 151
Theater of military action (operation) (TVD) (*see also:* Theater of strategic military action) 20, 133, 136, 141, 202, 225
Theater of strategic military action (TSMA) (*see also:* Theater of military action) 25, 26, 32, 33, 39, 40–2
Theater warfare 1, 21–2, 25–6, 33, 34–40, 44, 252
  role of nuclear weapons in 34–40, 140
Third World 2, 198, 252
  and France 117, 120, 122
  national liberation movements 74
  and Soviet Union 73, 122, 130
  and United States 80, 122, 252
Tindemans, L. 207
Tornado 136, 140, 150, 226
Treaty on the Non-Proliferation of Nuclear Weapons 167, 195, 196, 197, 203, 228
Trident (*see also:* Ballistic missile submarine; United Kingdom) 130, 131, 135, 138–40, 144, 153, 154
Trukhanovskii, V. G. 144, 155
TSMA *see* Theater of strategic military action
Tsoppi, Viktor 68
Tsvetkov, A. 227
Turkey 91, 205

TVD *see* Theater of military action
Tyushkevich, S. A. 34

United Kingdom (*see also:* Atlanticism; British Army on the Rhine; Conservative party; Europeanization; Labour party; Limited nuclear war option; Nimrod; Nuclear deterrence; Nuclear war; Peace movement; Social-Democratic party–Liberal alliance; Suez syndrome; Thatcher, Margaret; West European defense cooperation) 10–13, 27, 119, 120, 121, 127–58, 205, 223–4
  air force (*see also:* European Fighter Aircraft; Jaguar; Tornado; Vulcan bombers) 133, 135–7, 138, 140, 141
  arms sales 150
  army (*see also:* British Army on the Rhine) 133, 134, 137, 141
  bases on British soil 129, 131, 141, 142, 145, 158
  bridge between United States and Western Europe 129, 144, 146, 147, 158
  conventional forces and operations 133, 134, 136, 137, 140, 148, 157–8
  defense spending 136, 147, 158
  domestic politics 152
  force structure 127, 133–42
  military strategy 133–42
  navy (*see also:* Ballistic missile submarine; Chevaline; Falklands; Leander; Polaris; Trident) 133, 134–5, 137, 140–1, 253
  nuclear cooperation with United States 10–11, 12, 115, 129, 130, 144, 149, 253
  nuclear forces 10, 11, 12, 20, 21, 67, 68, 69, 112, 115, 121, 129, 131, 132, 133, 134, 135, 136, 137–41, 144, 145, 149, 152, 153, 154, 155, 156, 157, 158, 203, 205, 207, 228, 253, 254
  out-of-area 3, 117, 134
  role in NATO, Western Europe 11, 13, 80, 127, 128, 129, 133–42, 143, 146–8, 206
  security debate 127, 142–3, 151–7, 158
  security policy 127–58, 185, 218, 253
  and Soviet Union (*see also:* Conservative party; Labour party) 10–13, 66, 127–33, 143, 154, 155–7, 192
  Soviet wartime strategy toward 20–1
  (special) relationship with United States 10–11, 15, 20, 127, 128–9, 130–1, 132, 139, 142, 143–6, 147, 154, 155–6, 157, 158
United Nations 77, 102, 165, 166, 195
  Gorbachev speech to (7 December 1988) 17, 82
United States 11, 27, 37, 58, 106, 107, 108, 115, 121, 144, 147, 148, 156, 174, 193, 199, 203
  conventional forces 141, 145, 188, 190, 206, 254
  influence in Eastern Europe 3, 59
  influence in Western Europe 3, 5, 51, 59, 66, 74, 106, 111, 118, 146, 194, 207
  military policy 118, 122, 140, 185, 198, 202–3
  navy 117, 135, 173, 188
  nuclear policy (*see also:* Limited nuclear war option; North Atlantic Treaty Organization; Nuclear war; Theater warfare) 144, 169, 252
  nuclear weapons 3, 4, 6, 12, 17, 21, 27, 34, 69, 110, 137, 141, 145, 146, 149, 153, 155, 156, 188, 190, 191, 192, 202, 206, 208, 253, 254
  out-of-area 10, 117, 252
  presence in Western Europe 7, 8, 16, 17, 21, 59, 84, 188, 194, 225, 230, 231–2, 236, 251, 253, 254
  (reliability of) nuclear guarantee 7, 11, 12, 110, 119, 145, 149, 169, 193, 204, 252
  and Soviet Union 52, 53, 57, 58, 60, 61–2, 65–8, 98, 99, 132, 197, 233
Ural Mountains 95
Utkin, A. I. 8–9, 118, 135

van Oudenaren, John 62
Versailles 186
Vertical escalation 42
Vietnam 202
Vladimirov, O. 59
Voennaya mysl' 29, 46
Voigt, Karsten 245, 246
Volodin, I. 116
von Weizsaecker, Richard 222, 238
Vorneverteidigung see Forward defense
Vorob'yev, S. I. 46
Voronkov, L. S. 170, 171–2
Vorontsov, G. A. 143, 144
Voroshilov General Staff Academy lectures 20, 28, 31, 34, 39
Vulcan bombers 139

Waigl, Hans 244
Warsaw Pact 29, 41, 52, 53, 73, 85, 86, 90, 94, 102, 136, 138, 190, 193, 196, 197, 201, 202, 203
  cohesion 51, 91, 93, 94, 96, 98
  "Combined Armed Forces and Wartime Command Organs" agreement 41
  declaration on military doctrine and conventional arms control 17–18
  mobilization 41
  ties with NATO 71, 73
Wehrmacht 201

West European defense cooperation (*see also:* European fighter aircraft; European nuclear force; Eurogroup; Independent European Program Group; Western European Union) 3, 8, 10, 11–13, 17, 62, 64, 81, 107, 116–17, 119–20, 130, 133, 136–7, 141, 142, 146–7, 148–51, 204–9, 218, 226, 230, 251
   Franco-British 8, 10, 119–20, 141, 150, 151, 205
   Franco-German 3, 5, 8, 10, 17, 64, 79, 81, 108, 110, 119, 120, 204, 208, 225
   Soviet reaction to 4, 8, 10, 17, 62, 64, 117, 133, 208, 230
West European economic cooperation (*see also:* European Economic Community) 2, 3, 64, 108, 230, 251
   Soviet reaction to 2, 17, 64, 230
West European integration *see* West European defense cooperation; West European economic cooperation
Western Alliance *see* North Atlantic Treaty Organization
Western Europe (*see also:* Arms control; Economic activity; Europeanization; Gas pipeline; North Atlantic Treaty Organization; Nuclear-free zones; Power center; Third World; West European defense cooperation; West European economic cooperation)
   independence of 9, 51, 71, 79, 80, 106, 109, 120, 155
   intra-West European relations 5, 8, 10, 62, 63, 84, 107, 120, 122, 251, 252
   role of small- and medium-sized countries 98, 99, 163
   security relations with the United States *see* North Atlantic Treaty Organization; Nuclear hostage
Western European Union (WEU) (*see also:* West European defense cooperation) 10, 12–13, 16, 77, 110, 119, 120, 205–6, 208, 218
West German–Soviet Commission for Scientific and Technological Cooperation 234
West Germany *see* Federal Republic of Germany
Westland affair 130, 154
Westpolitik 97
WEU *see* Western European Union
Wilson, Harold 128, 129
Wimmer, Willy 209
Woerner, Manfred 228
World Bank 77
Worldnet 236
World War II 42, 72, 127, 183, 210–11, 227, 233
   relevance to today 18–19

Yakovlev, Aleksandr 56, 83, 232
Yazov, D. T. 32
Yeltsin, Boris 72
Yugoslavia 184

Zagladin, Vadim 61, 232, 235
Zarubezhnoe voennoe obozrenie 163
Zero option *see* Arms control
Zhukov, Yurii 64–5
Zhurkin, V. 211

For Product Safety Concerns and Information please contact our EU
representative GPSR@taylorandfrancis.com
Taylor & Francis Verlag GmbH, Kaufingerstraße 24, 80331 München, Germany

www.ingramcontent.com/pod-product-compliance
Lightning Source LLC
Chambersburg PA
CBHW071812300426
44116CB00009B/1285